Biogeography and Biodiversity of Western Atlantic Mollusks

Biogeography and Biodiversity of Western Atlantic Mollusks

Edward J. Petuch

Department of Geosciences, Florida Atlantic University

Photography by **Dennis Sargent**

CRC Press
Taylor & Francis Group
Boca Raton London New York

CRC Press is an imprint of the
Taylor & Francis Group, an **informa** business

CRC Press
Taylor & Francis Group
6000 Broken Sound Parkway NW, Suite 300
Boca Raton, FL 33487-2742

First issued in paperback 2017

© 2013 by Taylor & Francis Group, LLC
CRC Press is an imprint of Taylor & Francis Group, an Informa business

No claim to original U.S. Government works

ISBN-13: 978-1-4665-7979-8 (hbk)
ISBN-13: 978-1-138-03375-7 (pbk)

Library of Congress Cataloging-in-Publication Data

Petuch, Edward J.
   Biogeography and biodiversity of western Atlantic mollusks / Edward J. Petuch.
   pages cm
   Summary: "Detailing three marine molluscan faunal provinces (Carolinian, Caribbean, and Brazilian), this volume examines the distribution and evolutionary patterns of the marine mollusks of the tropical and subtropical regions of the western Atlantic Ocean. It also covers the paleobiogeography of the tropical Americas, discussing the ancestral biogeographical regions that gave rise to the recent provincial and subprovincial arrangements. These biogeographical patterns are essential for understanding the evolution of modern continental shelf marine faunas, as they reflect the environmental and climatological histories of the entire oceanic basin. The book's color plates make it a highly desirable resource for marine biologists and malacologists"-- Provided by publisher.
   Includes bibliographical references and index.
   ISBN 978-1-4665-7979-8 (hardback)
   1. Mollusks--Atlantic Ocean--Geographical distribution. 2. Mollusks, Fossil--Atlantic Ocean. 3. Paleobiology--Atlantic Ocean. I. Title.

QL408.2.P465 2013
594--dc23
                                                2013002110

Visit the Taylor & Francis Web site at
http://www.taylorandfrancis.com

and the CRC Press Web site at
http://www.crcpress.com

To my wife, Linda J. Petuch, and my children, Eric, Brian, and Jennifer, and to my scientific heroes, Rachel Carson and William Healey Dall

Contents

Foreword

The past several decades have witnessed an increased sense of urgency for describing and documenting the biodiversity of the planet. Entire new faunas have been discovered at hydrothermal vents, and vast numbers of new species have been, and continue to be, discovered from regions that were previously well studied. The addition of molecular techniques to the systematist's toolbox has shown clearly that what long had been regarded as well-known single species actually represent previously unknown species complexes composed of many new species.

Mollusks make up the most diverse phylum of marine organisms and are an ideal group for the study of patterns and processes that govern both how organisms evolve and how they are distributed throughout the marine realm. Interactions between the physiological requirements of the animals and physical parameters that change on time scales ranging from seasonal to geological have created a mosaic of regions, each with a characteristic habitat populated by a distinctive community of animals. The tropical western Atlantic has been an exceptionally dynamic area, both tectonically and ecologically, during much of the Neogene, and as a result hosts a rich molluscan fauna that is distributed over a broad array of distinctive regions.

Professor Petuch draws upon an extraordinary wealth of personal experience and many decades of field work studying both Recent and fossil mollusks throughout the western Atlantic, and has produced a prolific body of publications on these faunas. In the present work, he traces the development of a biogeographic framework for the temperate and tropical western Atlantic from its foundations in the mid 19th century through both qualitative and quantitative phases. Continuing this research, Petuch defines a series of molluscan faunal provinces and subprovinces based on his Provincial Combined Index, a novel refinement of Valentine's 50% rule. Using this concise quantitative model, he partitions the modern molluscan fauna of the tropical western Atlantic region into three provinces and fifteen subprovinces. Each is discussed in detail, accompanied by maps and excellent illustrations of index species, including *in situ* images of live animals and characteristic habitats. Provinciatones, primary and secondary relict pockets, and paraprovincialism are defined and incorporated into his discussions of individual subprovinces.

The use of 10 gastropod families and subfamilies as Provincial Index Taxa, with calculations of Provincial Combined Indices based on a summation of the relative endemism of each of these taxa, provides convenient and quantifiable approximations that are derived from a subset of taxa that serve as a proxy of total diversity. These indices have the advantage of being scalable as well as adaptable. Data from any number of additional families may be added to broaden the scale of the study. Such indices may also be used to explore and compare patterns among different phyla, between groups with varying forms of larval development, or inhabiting different bathymetric ranges within the same geographic region.

The author is to be commended for clearly and succinctly defining a useful tool for quantifying faunal distinctions among geographic regions. This methodology can also be used to produce a series of testable hypotheses that will serve both as a foundation and as a point of departure for additional research into the effects of geography and ecology on the evolution and diversification of faunas.

M. G. Harasewych, PhD
National Museum of Natural History
Smithsonian Institution

Acknowledgments

I would like to thank the following for their invaluable assistance in the production of this book. For the loan or donation of specimens photographed in this book, I thank the following: Richard Goldberg, Columbia, Maryland; Randy Allamand, Sebring, Florida; Albert Deynzer, Beverly Deynzer, and Neal Deynzer, Sanibel, Florida; Lyle Therriault, Concord, North Carolina; Thomas and Paula Honker, Delray Beach, Florida; Everett Long, Cedar Point, North Carolina; Dr. John Whicher, Somerset, England; Dr. Anton Oleinik, Florida Atlantic University; Jeffrey Whyman, Boynton Beach, Florida; Marcus and Jose Coltro, São Paulo, Brazil; Domingos Afonso Jorio, Guarapari, Brazil; Robert Pace, Miami, Florida; Kevan Sunderland, Plantation Key, Florida; Carole Marshall, West Palm Beach, Florida; Dr. Kenneth Keaton, Florida Atlantic University, and Wendy Keaton, Lauderhill, Florida; Andre Poremski, Washington, DC; Frank Frumar, Kirkwood, Missouri, and Cudjoe Key, Florida; Andrew Dickson, Miami, Florida; Eddie Matchett, Okeechobee, Florida; Clifford Swearingen, Fort Lauderdale, Florida; Nicolas Mazzoli, Pine Island, Florida; Dr. Luis Vela, director, Museo del Mar Mexico, Merida, Mexico; Brian Morgan, Cudjoe Key, Florida; Francis Hennequin, France; Bruno Besse, France; the late Leonard Hill, Miami, Florida; Dr. M. G. Harasewych, National Museum of Natural History, Smithsonian Institution, Washington, DC; and Jonnie Kuiper, Veendam, the Netherlands.

For helping with the collection of specimens on Glover's Atoll, Belize, I thank the following: Charles ("Chuck") and Linda Coons, Camano Island, Washington; Dennis Moore and Penelope ("Penny") Moore, Prospect, Oregon, and Loreto, Mexico; and John Clites, Brazil.

For help in collecting specimens in Amuay Bay, Venezuela, and Carriacou Island, Grenadines, I thank Dr. M. G. Harasewych, Smithsonian Institution, Washington, DC. I also thank Linda, Jenny, Eric, and Brian Petuch for assistance in collecting specimens on Eleuthera Island, Bahamas. Special thanks go to Dr. Alan Kohn, University of Washington, for allowing me to use his photographs of cone shell type specimens; to Andre Poremski, Washington, DC, for the photographs of his specimens of Bahamian, Grenadian, and Nicaraguan subprovince cone shells; to Dr. Anton Oleinik, Florida Atlantic University, for photographing the holotypes of the new species illustrated throughout the book; to William Bennight, North Carolina, for the pictures of living Florida Keys cone shells; and to Dennis Sargent, Mount Dora, Florida, for taking the beautiful photographs and setting up the plates of the shells illustrated in this book.

For reviewing the manuscript and offering many helpful suggestions, I give special thanks to the following: Dr. Gabriela Raybaudi Massilia, Roma Tre University, Rome, Italy; Dr. Manuel Tenorio, University of Cadiz, Spain; Dr. M. G. Harasewych, Smithsonian Institution, Washington, DC; John K. Tucker, Illinois Natural History Survey; Andre Poremski, Washington, DC; Marcus Coltro and Jose Coltro, São Paulo, Brazil; and Pierre Recourt, Egmond aan Zee, the Netherlands. For technical assistance and help in assembling this book, I give special thanks to John Sulzycki, senior editor; Jill Jurgensen, senior project coordinator; and Tara Nieuwesteeg, production editor, at CRC Press.

Introduction: American molluscan faunas in time and space

The science of marine biogeography, the study of the spatial distribution of organisms in the world's oceans, is one of the most fascinating branches of marine ecology. The distributions of molluscan faunas, in particular, were first observed by biologists over 150 years ago, and this book is meant to be a continuation of that pioneer research. As noticed by the early malacologists, three ecological factors limit the latitudinal distribution of marine mollusks: water temperature, salinity, and substrate type. Since most marine mollusks are physiologically constrained by both limited osmoregulatory abilities and temperature-sensitive enzyme systems, their distributions are arranged by latitude along the Neritic and Bathyal Zones of the continental margins. The spatial arrangement of malacofaunas, as outlined in this book, can be considered a proxy for western Atlantic oceanographic conditions and can be used as a tool for determining patterns of global climate change.

The ecological limitations and distributions of living mollusks are also powerful tools for the study of ancient marine environments. The latitudinal arrangement of Cenozoic fossil mollusks, particularly those from well-studied tropical families, is a precise reflection of paleotemperature gradients and ancient marine climates. Likewise, the salinity tolerances and osmoregulatory abilities of tropical families can also be used to determine the chemical and physiographic characteristics of ancient water masses. As will be seen in the living molluscan assemblages described later in this book, the vast majority of gastropod mollusks are also ecologically and morphologically tied to very specific substrate types. In the fossil record, these preferences for rocky, sandy, coralline, or muddy substrates can be used to give insight into past depositional environments. Because of the consistent ecological preferences seen in gastropods over the past 65 million years, marine biogeographers have been able to piece together a detailed history of the evolution of modern biotic patterns and shifting latitudinal boundaries. Recent biogeographical regions, then, should be viewed in the context of the constantly changing oceanographic parameters that have dominated the past epochs of the Cenozoic Era. Instead of a static, fixed system of biogeographical areas, the tropical western Atlantic is now known to be a dynamic region that has undergone rapidly occurring extinctions and evolutionary explosions, producing one of the richest molluscan faunas found anywhere on Earth.

As important as temperature, salinity, and substrate are in determining distributional patterns, sea level change (eustatic fluctuations) is a larger, overriding factor that supersedes those limiting factors. Occurring in regular cycles throughout geological time, primarily due to the building up and melting of continental glaciers, these fluctuations caused large areas of the western Atlantic continental shelves and offshore banks either to become completely emergent (during eustatic lows) or to be submerged under deeper water (during eustatic highs). During extended eustatic lows, some with sea levels dropping to

100–200 m below present level, spatial configurations were greatly diminished. This was particularly damaging on shallow banks and carbonate platforms. Instead of having a wide, diverse set of habitats across the banks and platforms, the inhabitable areas shrank to thin peripheral strips during times of low sea levels. In the time of a eustatic low stand, many species became extinct and others were separated into small, ecologically stressed populations. These sequestered faunules subsequently underwent rapid speciation due to genetic isolation. This process was reversed during eustatic highs, when the entire bank or platform was again submerged. The surviving species repopulated the newly available habitats and, again, underwent waves of speciation (the *founder effect*). The cycles of eustatic fluctuations, then, can be seen to act as forcing vectors for the evolution of new taxa. Many of the biogeographical patterns seen today in the western Atlantic are the result of eustatic fluctuations, ecological stress, and rapid evolutionary selection during the late Pleistocene Epoch. This multistage catastrophic process will be discussed later in this book.

On a personal note, I have always found molluscan biogeography to be one of the most intellectually satisfying branches of marine ecology. This is primarily due to the synthetic nature of the science and its inherent complexity. Any in-depth marine biogeographic study involves aspects of physical and chemical oceanography, systematic and evolutionary biology, historical geology, tectonics, paleontology, climatology, geomorphology, and physical geography. Each one of these disciplines, by itself, is a fascinating field of study, but the fusion of all of them into a single synthetic body of knowledge is aesthetically beautiful, rivaling elegant mathematical proofs. I gained much of the information used in this book from over 40 years of intensive field work: from diving, from working on shrimp and scallop boats, from cruises on oceanographic research vessels, and from extensive shore collecting around the western Atlantic. I have also had the opportunity to witness these biogeographical patterns firsthand by traveling and collecting mollusks in 14 countries in the Americas, from the southeastern United States to Uruguay. For me, some of the most exciting and rewarding research localities were on unstudied offshore archipelagos, such as Glover's Atoll off Belize, Los Roques Atoll off Venezuela, and the Abrolhos Islands off Brazil. Being one of the few people to ever collect mollusks on these isolated island groups, I was able to directly observe patterns of island endemism and, as a bonus, discovered and named many new endemic species. In the following chapters, I illustrate some of the more exotic and poorly known areas that were visited during these research trips. Hopefully, the locality photographs, along with the illustrations of over 300 species of rare and seldom seen shells, will underscore the intense beauty that is uniquely distributed across the subtropical and tropical areas of the western Atlantic Ocean.

Edward J. Petuch

About the author

Edward J. Petuch was raised in a Navy family, and spent many of his childhood years collecting living and fossil shells in such varied localities as California, Puerto Rico, Chesapeake Bay, and Wisconsin. His early interests in malacology and marine biology eventually led to BA (1972) and MS (1975) degrees in zoology from the University of Wisconsin–Milwaukee. While in Wisconsin, his thesis work concentrated on the molluscan biogeography of West Africa. There, he collected mollusks and traveled extensively in Morocco, the Canary Islands, Western Sahara, Senegal, Gambia, Sierra Leone, Ivory Coast, and the Cameroons.

Continuing his education, Dr. Petuch studied marine biogeography and malacology under Gilbert Voss and Donald Moore at the Rosenstiel School of Marine and Atmospheric Sciences at the University of Miami. During that time, his dissertation work involved intensive collecting and working on shrimp boats in Colombia, Venezuela, Barbados, Grenadines, and Brazil. After receiving his PhD in oceanography in 1980, Dr. Petuch undertook two years of postdoctoral research on paleoecology and biogeography with Geerat Vermeij at the University of Maryland. He also held a research associateship with the Department of Paleobiology at the National Museum of Natural History at the Smithsonian Institution (under the sponsorship of Thomas Waller) and conducted intensive fieldwork on the Plio-Pleistocene fossil beds of Florida and the Miocene of Maryland.

Dr. Petuch has collected fossil and living mollusks in Australia, Papua-New Guinea, the Fiji Islands, French Polynesia, Japan, the Mediterranean coast of Europe, the Bahamas, Mexico, Belize, and Uruguay. This research has led to the publication of over 100 papers. His 14 previous books (including *Cenozoic Seas: The View from Eastern North America*, *The Geology of the Everglades and Adjacent Areas*, *Molluscan Paleontology of the Chesapeake Miocene*, *Atlas of the Living Olive Shells of the World*, *New Caribbean Molluscan Faunas*, and *Atlas of Florida Fossil Shells*) are well-known research texts within the malacological and paleontological communities.

Presently, Dr. Petuch is professor of geology in the Department of Geosciences at Florida Atlantic University in Boca Raton, where he teaches courses on oceanography, paleontology, and physical geology.

chapter one

The molluscan provincial concept in the tropical western Atlantic

For over 150 years, marine biologists, malacologists, and oceanographers have recognized that the shallow water marine molluscan faunas of the world are distributed in a pattern of distinct, geographically definable areas. Intensive shell collecting by nineteenth-century naturalists in the English, Spanish, French, and Dutch tropical island colonies resulted in the discovery of an overwhelming number of spectacular new species and genera. These new taxa supplied the first evidence for the existence of larger biogeographical patterns. As more data poured into Europe in the late 1800s and, later in the United States in the early 1900s, these geographical patterns became better defined. Building on this new pool of information, the European and American pioneer molluscan biogeographers, in recognizing latitudinal distributions, were actually planting the seeds for the development of the science of marine ecology. The evolution of the thought processes that resulted in our present understanding of western Atlantic marine biogeography is as fascinating as the subject itself.

History of molluscan biogeographic research in the western Atlantic

A framework for the biogeography of the tropical western Atlantic was first proposed in two contemporaneous books, both published in 1856. The first of these, *The Physical Atlas of Natural Phenomena* by Alexander Johnston, divided the world into 25 provinces within a series of nine horizontal, latitudinally arranged homoizoic belts. This is generally considered to be the first comprehensive book ever published on marine biogeography and was the first to use the word *province* as a biogeographic unit. In his *Manual of Conchology*, which was published at the same time (1856), Samuel Woodward described 25 worldwide marine molluscan faunal provinces, with three encompassing the area covered in this book. Based on preliminary data published by Edward Forbes (considered to be the father of modern marine biology), Woodward's three western Atlantic provinces are the (1) *Transatlantic Province* (Cape Cod, Massachusetts, to eastern Florida), with two subdivisions, the *Virginian* (Cape Cod to Cape Hatteras) and the *Carolinian* (Cape Hatteras to eastern Florida); (2) the *Caribbean Province* (the Gulf of Mexico, Caribbean Basin, and the Antilles, south to Rio de Janeiro, Brazil); and (3) the *Patagonian Province* (from Santa Catarina State, Brazil, to Punta Melo, Argentina and the Falkland Islands). Both Johnston's and Woodward's provinces were established on qualitative, not quantitative, aspects of the component molluscan faunas.

Although only qualitative in nature, Woodward's scheme of 25 separate worldwide molluscan provinces became the standard for marine biogeography throughout the nineteenth century and the first half of the twentieth century. This was reversed, however, in 1953, when Sven Ekman published his informative survey of worldwide marine faunas,

Zoogeography of the Sea. Although covering all types of marine organisms, not only the Mollusca, Ekman abandoned the provincial concept, preferring instead sets of regions and subregions. In Ekman's scheme, the area covered by this book would be included within his *Atlanto-East Pacific Region* and *Tropical and Subtropical American Subregion*. Following this concept of a single large tropical region in the western Atlantic, Germaine Warmke and R. Tucker Abbott (in *Caribbean Seashells*, 1961: 319) recognized a Caribbean Province, which extended from Cape Hatteras and Bermuda, throughout the southern Gulf of Mexico and the Caribbean Basin, the Antilles, and the entire coast of Brazil south to Cabo Frio. The northern Gulf of Mexico and the coastal areas of Georgia and the Carolinas were not included in their Caribbean Province. Both Ekman's and Warmke and Abbott's biogeographical schemes are now considered to be overgeneralized and, intrinsically, do not offer the level of resolution needed for the recognition of localized species radiations and island endemism.

The concept of broad, generalized faunal regions was challenged in the 1970s, primarily by the works of James Valentine. Expanding on Woodward's original marine molluscan province concept, Valentine (1973: 337) was the first to offer a quantitative definition of a faunal province. His scheme involved the use of a *50% Rule*, where at least one-half of the species living within the province must be endemic. Valentine also demonstrated the use of cluster analyses, Jaccard's coefficient, and other statistical methods in the determination of provincial and subprovincial boundaries. Applying these techniques to the known molluscan taxonomic data of the time, he defined four distinct provinces for the area covered by this book: the Carolinian Province (Cape Hatteras to eastern Florida); the Gulf Province (southern and western Florida, the entire Gulf of Mexico to Yucatan, Mexico); the Caribbean Province (Cuba, the West Indies, the Caribbean Basin, northern South America south to Cabo Frio, Brazil); and the Patagonian Province (Cabo Frio south to the Golfo San Matias, Argentina). Although retaining three of Woodward's provincial names (Carolinian, Caribbean, and Patagonian), Valentine redefined their boundaries and added a fourth new province, the Gulf. This analytical study, although now outdated, was the first to pin down provincial boundaries by statistics and was also the first to use plate tectonics and physiological restrictions (limiting factors) to explain provincial distributions.

Several years after Valentine's monumental work, Geerat Vermeij defined two broad provinces in his marine ecological study titled *Biogeography and Adaptation* (1978): (1) a *Tropical Western Atlantic Province* (from Palm Beach and Bermuda, the southern Gulf of Mexico, the Caribbean Basin, and south to Brazil); and (2) a *Warm-Temperate Northwest Atlantic Province* (Cape Hatteras, North Carolina, south to Palm Beach, Florida). These broad faunal regions were based primarily on latitudinal patterns of predation on mollusks and not on their actual molluscan faunal compositions. Vermeij's work, however, inspired me to conduct intensive fieldwork in the tropical Americas in the late 1970s and early 1980s. During this time, I concentrated my explorations on the southern Caribbean, particularly the Grenadines, Venezuela, Colombia, and Barbados and the entire coastline of Brazil. From these then unexplored areas, large numbers of new species and genera were collected, giving new insight into the biogeographical patterns of the Caribbean region and South America. All of these important new taxa and their accompanying biogeographical patterns were summarized in one large compendium titled *New Caribbean Molluscan Faunas* (Petuch, 1987). Building on this new taxonomic data, I later redefined the provincial arrangement of the tropical western Atlantic, recognizing Woodward's Carolinian and Caribbean Provinces and adding a third new province, the *Brazilian Province* (Petuch, 1988: 166). In a later book, *Cenozoic Seas: The View from Eastern North America* (Petuch, 2004: 56),

the Brazilian Province was further subdivided into four separate subprovinces. My inter-
pretations of the Carolinian, Caribbean, and Brazilian Provinces will be discussed in
detail in the following sections of this book.

Modifying the basic definitions of Woodward's and Valentine's Carolinian and
Caribbean Provinces and incorporating my Brazilian Province, John Briggs offered a new
provincial arrangement in his book titled *Global Biogeography* (1995). Building upon his earlier work, *Marine Zoogeography* (1974), which emphasized ichthyofaunas, Briggs proposed a
series of three provinces and two regions for the tropical western Atlantic: the (1) *Carolinian
Region* (from Cape Hatteras to northeastern Florida and the northern Gulf of Mexico), the
Caribbean Province (southern Florida, the southern Gulf of Mexico, coastal Central America,
and northern South America), and the *West Indian Province* (Bermuda, the Bahamas, Cuba,
and the Antilles); (2) the *Brazilian Province* (from the Orinoco River, Venezuela, to Rio de
Janeiro); and (3) the *Eastern South American Region* (Rio de Janeiro south to the Golfo San
Matias). Briggs considered the Carolinian and Eastern South American Regions to be
broad faunal transition zones and defined them as undifferentiable. In his scheme, Briggs
truncated the Brazilian Province and split the Caribbean area into two separate provinces,
essentially defining the continental coastal areas as being faunistically different from the
island archipelagos. Like Ekman and Vermeij, Briggs incorporated the biogeographical
data of many phyla and classes and attempted to define areas by their entire marine biota,
not just the molluscan faunas.

Probably the most unusual biogeographical scheme yet proposed was given by Juan
Diaz in his paper titled "Zoogeography of Marine Gastropods in the Southern Caribbean:
A New Look at Provinciality" (1995). In this short work, Diaz described several new prov-
inces and areas: the *Isthmian* (Panama to Santa Marta, Colombia); the *Samarian–Venezuelan*
(which is split into two parts, one extending from Santa Marta, Colombia, to the southern coast of the Goajira Peninsula, Colombia; and the other part extending from the
Golfo Triste, Venezuela, to northern Trinidad); and the *Goajira* (the northern coast of the
Goajira Peninsula, Colombia, the entire Gulf of Venezuela, and the Paraguaná Peninsula,
Venezuela). Diaz also defined two broad transitional areas: (1) the *Transitional Area toward
the Brazilian Province* (from southern Trinidad to Suriname); and (2) the *Transitional Area
to the Antillean Province* (northern Trinidad and the coralline islands off Venezuela). His
problematical split Samarian–Venezuelan Province, with the intervening Goajira Province
sandwiched in between, is now known to be simply an artifact of substrate type along
northern South America. The Samarian–Venezuelan actually represents areas where
muddy, organic-rich substrates dominate, while the Goajira represents an area where carbonate substrates dominate (shell hash and coralline algae). There are many eurytopic gastropods that span both provinces (such as *Voluta musica*; see Chapter 8 of this book) and a
detailed faunal analysis (see also Chapter 8) does not support this bifurcated system. Diaz
apparently confused widespread biotopes and their associated specialized molluscan faunas with actual provinces.

Interestingly enough, of all the marine biogeographical classification systems that
have been proposed over the past 150 years, only that of Woodward (1856) comes closest to
what the newest and most detailed data support. Because of his amazing scientific accu-
racy and foresight, Woodward truly deserves the title of father of molluscan biogeography.
In this book, I recognize three provinces within the tropical western Atlantic Region: the
Carolinian and Caribbean Provinces of Woodward (in modified form) and the Brazilian
Province that I proposed in 1988. These are discussed in Chapter 2.

Definition of the molluscan faunal province

As can be seen in the previous section, most of the biogeographical models were based entirely, or almost entirely, on qualitative data. Only Valentine (1973) proposed a formal definition of a province that involved an actual numerical index (the 50% rule). I restated Valentine's provincial definition (Petuch, 2004: 21) by pointing out that "two adjacent geographical areas can be considered separate molluscan provinces if at least 50% of the species-level taxa are endemic to each area. This mutual exclusivity includes the faunas of all possible analogous biotopes within the two provinces." Using this 50% rule as a standard for comparison, a higher resolution and far more accurate measure of provinciality is now possible.

Since this book deals with tropical molluscan provinces, the physical aspects of these biogeographical units must also be defined by quantitative data. Using water temperature as the primary parameter, what have been referred to as subtropical and tropical water conditions can be used to define *eutropical provinces* and *paratropical provinces* (Petuch, 2004: 21–22). Eutropical provinces are low-latitude, circumequatorial faunal regions where surface water temperatures do not fall below 20 degrees Celsius. These types of provinces contain molluscan faunas that are derived from high-tropical Eocene Tethys Sea lineages. The Caribbean Province (western Atlantic), Guinean Province (central West Africa), Panamic Province (Eastern Pacific), and Indo-Pacific Province (central Indian and Pacific Oceans) all fall within this category. Eutropical areas are characteristically bounded, in mirror-image fashion, by a pair of paratropical provinces. These higher latitude areas contain essentially subtropical or warm-temperate seas, where the surface water temperature regularly falls below 20 degrees Celsius in the winter but never drops below 10 degrees Celsius. Paratropical provinces typically contain mixed faunas, with Tethys Sea high tropical genera and species coexisting with cold water and boreal or antiboreal genera and species. Classic examples of paratropical provinces include the Californian Province (central California to the central part of Baja California, Mexico) and the Peruvian Province (northern Peru to northern Chile), which border the Eutropical Panamic Province; the Carolinian Province (Cape Hatteras, Florida, and the entire Gulf of Mexico) and Brazilian Province (Amazon River mouth to Uruguay), which border the Eutropical Caribbean Province; and the Lusitanian Province (southern Portugal to Mauretania, West Africa) and Cape Province (Namibia to Natal, South Africa), which border the eutropical Guinean Province.

Typically, Tethyan-derived eutropical provinces contain three worldwide environmental markers: (1) *turtle grass beds* (*Thalassia*), found in shallow lagoons; (2) *red mangrove forests* (*Rhizophora*), growing in intertidal and estuarine conditions; and (3) extensive *zonated coral reef complexes* composed of hermatypic scleractinian corals. In regard to this last criterion, extensive coral reefs are present only within the Caribbean and Indo-Pacific Provinces. The eutropical Guinean Province of West Africa contains extensive coral reefs only on the islands of São Thome, Principe, and Annobon, while the Panamic Province of western Central America and northwestern South America contains major coral complexes only around the Perlas Islands of the Bay of Panama and in isolated areas such as the southern tip of Baja California, the Nayarit coast and Tres Marias Islands of Mexico, and within the Golfo de Nicoya of Costa Rica.

Each of these three environmental markers has its own characteristic Tethyan-derived molluscan fauna, with the same genera and families being found in all high tropical areas around the world. Twenty-three gastropod families and subfamilies are typically associated with these three environments, and the presence of all of them indicates a eutropical province. These provincial indicators include the families Strombidae, Cypraeidae,

Ovulidae, Potamididae, Xenophoridae, Modulidae, Melongenidae, Cassidae, Bursidae, Personiidae, Turbinellidae, Costellariidae, Conidae, and Conilithidae and the subfamilies Muricinae, Muricopsinae, Ergalitaxinae, Fasciolariinae, Lyriinae, Moruminae, Olivinae, Cancellariinae, and Plesiotritoninae. The systematics of 10 of these wide-ranging families and subfamilies are now very well understood and well documented, and I have chosen these taxonomically more secure groups for quantitative analyses of provincial and sub-provincial boundaries. Referred to here as *provincial index taxa,* these include:

Families
    Modulidae
    Turbinellidae
    Conidae
    Conilithidae
Subfamilies
    Muricinae
    Fasciolariinae
    Lyriinae
    Olivinae
    Cancellariinae
    Plesiotritoninae

The genera and species contained in these 10 provincial index taxa are listed in Appendix 1. Each provincial index taxon yields a quantity, *T* (the *taxon index*), such that

$$T = \frac{n}{N}(100), \ T > 50$$

where *N* is the total number of species in the provincial index taxon, and *n* is number of endemic species; *T* must be greater than 50. A combined index, incorporating all 10 of the taxa indices yields a quantity *P*, the *provincial combined index*, such that

$$P = \sum_{n=1}^{10} \frac{Tn}{10}, \ P > 50$$

where *T*1 is the % endemism in the Modulidae; *T*2 is the % endemism in the Turbinellidae; *T*3 is the % endemism in the Conidae; *T*4 is the % endemism in the Conilithidae; *T*5 is the % endemism in the Muricinae; *T*6 is the % endemism in the Fasciolariinae; *T*7 is the % endemism in the Lyriinae; *T*8 is the % endemism in the Olivinae; *T*9 is the % endemism in the Cancellariinae; and *T*10 is the % endemism in the Plesiotritoninae. The provincial combined index must be greater than 50 for an area to be considered a full province. This combined index of the percentages of endemism in the provincial index taxa provides a higher resolution quantity that can be used for comparisons with adjacent biogeographical areas.

Unlike eutropical provinces, which contain a complete compliment of the 23 tropical index families and subfamilies, the paratropical provinces contain only a partial complement, with some, but not all, of these index groups being present. Occurring on the boundaries of high-tropical water conditions, paratropical provinces typically lack well-developed carbonate environments, along with their associated coral reefs, turtle grass beds, and red mangrove forests. Paratropical provinces also show stronger latitudinal differences than are seen in the more uniform eutropical provinces, with the higher latitudinal sections containing fewer examples of tropical index taxa than do the lower latitude sections. In the Recent Americas, a classic example of a paratropical province is seen in the Californian Province, which lacks coral reefs, turtle grass beds, and mangrove forests but where two of the provincial index taxa are represented by a single species: the Conilithidae (with *Californiconus californicus*) and Cypraeidae (with *Neobernaya spadicea*). Such cold-tolerant tropical derivatives coexist with classic cold-water groups such as *Euspira* (Naticidae), *Neptunea* (Neptuniidae), *Ophiodermella* (Turridae), and *Ceratostoma* (Muricidae-Ocenebrinae), producing a peculiar mixed molluscan fauna. The families Modulidae, Turbinellidae, and Conidae and the subfamilies Muricinae, Fasciolariine, Lyriinae, Olivinae, and Plesiotritoninae are absent from the Californian province. As Vermeij (1978) pointed out, the cold-tolerant (cryophilic) offshoots of tropical families are usually generalized feeders with weak shell architecture, demonstrating that their invasion into higher and lower latitudes may be the result of avoidance of predation and competition. In this book, both eutropical and paratropical provinces are considered to make up the Tropical Western Atlantic Region.

Definition of the molluscan subprovince

Geographically widespread provinces typically contain areas with multiple endemic species radiations and localized evolutionary centers. These special endemic organisms are almost always species that are nonvagile, with direct-development larvae that do not have planktotrophic stages. The larva remains within the egg capsule and hatches into a crawl-away miniature adult, never dispersing more than a few miles from where it was born. These types of nonvagile mollusks are highly susceptible to genetic isolation and rapidly undergo speciation, particularly along marginal areas where seasonal climate changes are more extreme. Because of this, multiple areas within a province may develop their own characteristic faunas and become distinctive enough to be separable from the rest of the province.

Geographically definable faunal subdivisions within a province, containing localized centers of allopatric speciation, are referred to as subprovinces. Originally (Petuch, 2004: 21), subprovincial areas were defined by a *30% Rule*, where at least 30% of the species of the 10 key index families and subfamilies were endemic to the area. In light of more detailed data on tropical gastropod faunas, I have now modified this percentage of endemism into a *25% Rule* (one-half that of the province). The same basic algorithm that is used for provinces can be modified to yield *S* (the *subprovincial combined index*), such that

$$S = \sum_{n=1}^{10} \frac{tn}{10}, \; S > 25$$

where *t*1 is the % endemism in the Modulidae within the subprovince; *t*2 is the % endemism in the Turbinellidae within the subprovince; *t*3 is the % endemism in the Conidae within the subprovince; *t*4 is the % endemism in the Conilithidae; *t*5 is the % endemism in the Muricinae within the subprovince; *t*6 is the % endemism in the Fasciolariinae within the subprovince; *t*7 is the % endemism in the Lyriinae within the subprovince; *t*8 is the % endemism in the Olivinae within the subprovince; *t*9 is the % endemism in the Cancellariinae within the subprovince; and *t*10 is the % endemism in the Plesiotritoninae within the subprovince. The subprovincial combined index must be greater than 25 for an area to be considered a subprovince. Applying this rule to the provinces of the western Atlantic and using the data listed in the provincial index taxa section at the end of the book, 15 subprovinces are now known to exist, with five in the Carolinian Province, seven in the Caribbean Province, and three in the Brazilian Province. In context with the world-wide biogeographic classification scheme, these subprovinces include

Marine Biosphere
Benthonic Realm
Atlantic Neritic Subrealm
Tropical Western Atlantic Region
Carolinian Province
   Georgian Subprovince
   Floridian Subprovince
   Suwannean Subprovince
   Texan Subprovince
   Yucatanean Subprovince
Caribbean Province
   Bermudan Subprovince
   Bahamian Subprovince
   Antillean Subprovince
   Nicaraguan Subprovince
   Venezuelan Subprovince
   Grenadian Subprovince
   Surinamian Subprovince
Brazilian Province
   Cearaian Subprovince
   Bahian Subprovince
   Paulinian Subprovince

These provinces and subprovinces will be discussed in detail in the following sections and chapters. Using Olive shells (Family Olividae, Subfamily Olivinae) as biogeographical index taxa, a preliminary form of this hierarchical scheme was first proposed by Petuch and Sargent (1986). In light of new discoveries uncovered in the 1990s and early 2000s, the western Atlantic subprovincial nomenclatural scheme proposed in the 1986 book is now considered outdated and should be superseded by the scheme proposed in this book.

Provinciatones

Depending on the patterns and configurations of oceanic currents, the boundaries between two provinces are often blurred, containing wide, poorly demarcated areas of faunal overlap. These transition zones characteristically contain sympatric faunal elements from both parent provinces, often coexisting in unusual and novel molluscan assemblages. These broad areas of overlap, with shared faunal elements, also contain endemic species that are not found in either of the overlapping provinces but are unique to the boundaries of the transitional zone. This type of provincial overlap zone was referred to as a *provinciatone* (Petuch, 2004: 21) and is analogous to the smaller-scale ecotone (an overlap area between two ecosystems; Odum, 1971). Provinciatones characteristically contain three types of faunal components: warm-tolerant elements from the higher latitude province; cold-tolerant elements from the lower latitude province; and *provinciatonal endemics* that are restricted to the boundaries of the provinciatone.

In the tropical western Atlantic, two large provinciatones are known to exist: the *Palm Beach Provinciatone* (from approximately Fort Pierce to Dania Beach, Florida) and the *Uruguayan Provinciatone* (from approximately Rio Grande do Sul, Brazil, to Mar del Plata, Argentina). The Palm Beach Provinciatone represents the overlap of the subtropical Carolinian Province and the high tropical Caribbean Province, while the Uruguayan Provinciatone represents the overlap of the subtropical Brazilian Province and the cold-water Patagonian Province. Both provinciatones contain gastropod provinciatonal endemics, with those in the temperature-sensitive tropical family Conidae being of primary interest. Some of the cone shells that are restricted to the Palm Beach Provinciatone include *Dauciconus glicksteini*, *Kellyconus binghamae*, *Gradiconus patglicksteinae*, and *Tuckericonus flamingo* (illustrated and discussed in more detail in Chapter 3). For the Uruguayan Provinciatone, a single provinciatonal endemic cone shell, *Lamniconus carcellesi*, is known to range all the way to northern Argentina (see Chapter 11). This is amazingly far south for a member of the tropical family Conidae, and this indicates that the provinciatonal conditions existed there for a very long time, probably since the Pliocene (represented by the *Lamniconus platensis* species complex of the Camachoan Paleoprovince of the late Miocene-Pliocene; Petuch, 2004: 53; see the section on paleoprovinces at the end of this chapter). Other provinciatonal endemics from the Palm Beach and Uruguayan Provinciatones are given in Chapters 3 and 11.

Geographical heterochrony

Working with Colombian commercial shrimpers during my doctoral field research in the 1970s, I discovered the presence of "living fossil" (relictual) molluscan assemblages along the northern coast of South America (see Chapter 8 of this book). I later published on the unusual composition of the northern Colombian–Venezuelan offshore fauna (Petuch, 1979b, 1981a), pointing out the existence of many genera and species complexes that were thought to have become extinct by the late Pliocene Epoch but were found to be extant within this geographically small area. Some of these relict taxa included living members of the genera *Aphera* (Cancellariidae-Cancellariinae), *Conomitra* (Volutomitridae) and *Paraborsonia* (Turridae), and members of the *Tenorioconus consobrinus*, *Muracypraea henekeni*, *Sconsia laevigata*, and *Strombina caboblanquensis* species complexes, all of which are common in the late Miocene and Pliocene fossil beds of northern South America and Panamá. This discovery showed that the molluscan faunas of northern Colombia and Venezuela were archaic in nature and contained remnants of Mio-Pliocene southern Caribbean faunas that predate the closing of the Isthmus of Panamá, over 3 million years ago. These relictual communities were associated with wind-driven upwelling systems of cold, nutrient-rich water, simulating oceanographic conditions that were prevalent in the southern Caribbean before the closure of Panamá (Petuch, 1982a; Vermeij and Petuch, 1986). With the lack of habitat change for over 3 million years, the Colombian–Venezuelan faunas underwent an evolutionary stasis, retaining much of their original community structure.

During the time when the relictual assemblages of northern South America were first being studied in detail, other areas around the tropical western Atlantic were also found to harbor relictual faunas. Principal among these was the area around the Bay Islands of Honduras (Roatan, Utila, and Guanaja Islands) and the adjacent Honduran coast, where numerous other relictual taxa were discovered (Petuch, 1980a, 1981b, 1987). In this case, many of these relicts were previously known only from the Pliocene and early Pleistocene fossil beds of the southeastern United States and included the supposedly extinct genera *Pleioptygma* (Pleiopygmatidae) and *Cerithioclava* (Cerithiidae). Likewise, the Yucatan Peninsula of Mexico was also found to harbor many relictual taxa, some of which were known as fossils from the Bermont Formation (mid-Pleistocene) of southern Florida (Petuch, 1987, 1994, 2004; Petuch and Roberts, 2007). Some of these included *Melongena (Rexmela) bispinosa* (Melongenidae), *Fulguropsis* cf. *feldmanni* (Busyconidae), and *Vokesimurex anniae* (Muricidae-Muricinae). The outer neritic region off the western coast of Florida was also found to house numerous Pliocene relictual elements, many of which were known only from the Pliocene Tamiami Formation of southern Florida. Some of these included living members of the *Lindafulgur lindajoyceae* species complex (*L. lyonsi*; Busyconidae), the *Cancellaria floridana* species complex (*C. richardpetiti*; Cancellariidae), and the *Torculoidella ochlockoneensis* species complex (*T. lindae*; Turritellidae). I later described and illustrated many of the Colombian, Venezuelan, Honduran, Yucatanean, and western Floridian relictual taxa in a series of publications, the largest and most comprehensive being *New Caribbean Molluscan Faunas* (Petuch, 1987) and *Neogene History of Tropical American Mollusks* (Petuch, 1988).

All of these isolated and disjunct faunal areas within the western Atlantic were referred to as *relict pockets* (Petuch, 1982a), and two distinct types were distinguished: (1) *primary relict pockets*, where most of the relictual components are the same genus and species as those found in the fossil record; and (2) *secondary relict pockets*, where most of the relictual components belong to supposedly extinct genera but have evolved into species different from those found in the fossil record. The Yucatan area, with its intact Bermont Formation fauna (mid-Pleistocene), represents a primary relict pocket, while western Florida and northern Colombia, with their living Pliocene genera and descendant species, represent secondary relict pockets. Other relictual faunas have subsequently been discovered, including secondary pockets off Barbados and along northern Brazil. These all demonstrate that the western Atlantic molluscan faunas are evolving at different rates, with some having changed little since the Pliocene and Pleistocene and others having undergone rapid evolution, resulting in completely new post-Pleistocene assemblages. This type of contemporaneous coexistence of archaic and modern faunas, arranged in a "patchwork quilt" pattern, is referred to as *geographical heterochrony* (Petuch, 1982a). The distribution of these relict pockets, along with the patterns of geographical heterochrony found in the western Atlantic provinces, will be discussed throughout the following chapters.

Submergence and endemic bathyal faunas

In the tropical western Atlantic, the distribution of molluscan faunas in the upper part of the bathyal zone (upper continental slope from 200 to 500 m) mimics the biogeographical patterns seen in the neritic and littoral zones (intertidal to 200 m depths). This pattern, referred to here as *submergence*, implies that these deeper water offshore faunas are derived, evolutionarily, from the adjacent shallow water communities. Although the water temperatures along the upper bathyal zone are relatively uniform around the tropical western Atlantic (with notable exceptions discussed in the sections on the Bahamian and Grenadian Subprovinces), the distributions of molluscan faunas are not, instead conforming to the boundaries of the subprovince. Because of this, the subprovincial areas of the western Atlantic contain two suites of geographically confined species: one in the shallow neritic zone (0–200 m) and one in the deep-water upper bathyal zone (200–500 m).

In many cases, the upper bathyal faunas contain genera and species groups that were, during the Pliocene and Pleistocene, living in shallow intertidal depths (Petuch, 1982a, 1994). This type of submergence is particularly characteristic of the subprovinces of the Carolinian Province in the Gulf of Mexico, where Pliocene shallow water (0–5 m depths) groups such as *Busycoarctum* and *Lindafulgur* (both Busyconidae), *Scaphella* (Volutidae), and *Heilprinia* (Fasciolariidae) now live in deep-water offshore areas. These genera are abundant as fossils in the Pliocene Tamiami Formation and early Pleistocene Caloosahatchee Formation of southern Florida, both of which were deposited in shallow lagoonal environments inside the enclosed Okeechoban Sea of the Everglades area (see Petuch, 2004; Petuch and Roberts, 2007 for descriptions of the Okeechoban Sea environments and molluscan faunas). As sea levels dropped precipitously at the end of the Piacenzian Stage of the Pliocene and later at the end of the Calabrian Stage of the Pleistocene, these shallow-water elements followed the littoral environments down the continental shelf. When sea levels again began to rise during the warm interglacial stages of the Pleistocene and Holocene, these cold-adapted relictual taxa remained in the cooler, deeper water areas, having become established within permanent bathymetric refugia.

Other patterns of submergence involving derivations of Neogene shallow water faunas are also known from the coasts of northern South America, Barbados, and the banks and seamounts off Honduras. These offshore faunas typically contain genera that are common as fossils in the shallow water facies of the Gatun Formation of Panama and the Punta Gavilan Formation of Venezuela. Some of these groups that have undergone submergence in this area of the Caribbean include *Aphera* (Cancellariidae), *Cotonopsis* and *Strombina* (Columbellidae), and the *Muracypraea cubaguensis* species complex (Cypraeidae; with living species such as *M. tristensis*). Endemic bathyal faunas and patterns of submergence will be discussed in more detail in the sections on deep-water faunas in the following chapters.

Western Atlantic paleoprovinces and paraprovincialism

The configurations and faunal compositions of the western Atlantic provinces and subprovinces reflect their oceanographic histories. The inception of the modern provincial arrangement dates to the Eocene–Oligocene boundary (36 million years BP), with the catastrophic impact of a giant asteroid (see Petuch and Roberts, 2007 for a discussion of the impact and its effects on the marine environment). Hitting what is today Chesapeake Bay, this huge asteroid produced numerous mega-tsunamis, which devastated the southeastern coast of the United States and Gulf of Mexico, and immense debris clouds, which plunged the world into a "nuclear winter" cold time. The rapid climatic degeneration produced a mass extinction in the molluscan faunas of the western Atlantic, with such common and characteristic Tethyan tropical genera as the immense horn shell *Campanile* (Campanilidae), the giant cowrie *Gisortia* (Cypraeidae), and the spiny volute *Athleta* (Volutidae) disappearing abruptly and never again occurring in the Americas.

When the marine climate once again returned to tropical conditions during the early Oligocene Epoch (Rupelian Stage), the survivors of the terminal Eocene extinction event had rearranged themselves into new biogeographical patterns. These reflected the changing current structures, water temperatures, and new coastline configurations that formed immediately after the impact. The new provinces and subprovinces that evolved during Rupelian time existed until the end of the Oligocene Epoch (Chattian Stage), and these became the evolutionary ancestors of all the modern western Atlantic biogeographical units. The Oligocene paleoprovinces were also the center of evolution for most of the living western Atlantic mollusks. Some of the western Atlantic endemics that first appeared within these paleoprovincial boundaries include the genera *Scaphella* (Volutidae; as *S. demissa*) and *Atlanticonus* (Conidae; as *A. kendrewi*), and the families Busyconidae and Pleioptygmatidae. The Oligocene American paleoprovinces and paleosubprovinces include

Tropical Western Atlantic Region—*Oligocene Epoch*
  Prototransmarian Paleoprovince (North Carolina to Nova Scotia)
  Antiguan Paleoprovince (North Carolina to Brazil)
    Hernandoan Paleosubprovince (North Carolina to Florida)
    Vicksburgian Paleosubprovince (Mississippi Valley and Gulf of Mexico)
    Alazanian Paleosubprovince (Mexico to Honduras)
    Guanican Paleosubprovince (Cuba and the West Indies)
    Bohioan Paleosubprovince (Honduras to northern Brazil)
  Pernambucan Paleoprovince (northern and central Brazil)

These Oligocene paleoprovinces and their molluscan faunas are discussed in detail in *Cenozoic Seas: The View From Eastern North America* (Petuch, 2004: 23–30).

The end of the Oligocene was also a time of severe climatic degeneration (asteroid impacts?), and this caused many of the characteristic older genera and species complexes to become extinct. By the beginning of the Miocene Epoch (Aquitanian Stage), however, the surviving faunal elements from these older paleoprovinces had begun to rearrange themselves into new biogeographical patterns. Many of the modern Carolinian, Caribbean, and Brazilian endemic genera evolved at this time, including *Pleioptygma* (Pleioptygmatidae), *Muracypraea* (Cypraeidae), and *Voluta* (Volutidae). The Carolinian Province endemic family Busyconidae also underwent a huge species radiation, resulting in the evolution of the living genera *Busycon, Lindafulgur,* and *Busycotypus.* The Miocene American paleoprovinces and paleosubprovinces include

Tropical Western Atlantic Region—*Miocene Epoch*
  Transmarian Paleoprovince (Nova Scotia to South Carolina)
    Sankatian Paleosubprovince (Nova Scotia to New Jersey)
    Calvertian Paleosubprovince (New Jersey to Virginia)
    Pungoian Paleosubprovince (Virginia to South Carolina)
  Baitoan Paleoprovince (South Carolina to northern Brazil)
    Onslowian Paleosubprovince (South Carolina to northern Florida)
    Chipolan Paleosubprovince (Florida to Texas)
    Agueguexquitean Paleosubprovince (Texas to Honduras)
    Anguillan Paleosubprovince (Cuba and the Antilles)
    Culebran Paleosubprovince (Honduras to Colombia)
    Cantaurean Paleosubprovince (Colombia to Suriname)
    Carriacouan Paleosubprovince (Lesser Antilles)
    Piraban Paleosubprovince (northern Brazil)
  Platensian Paleoprovince (central Brazil to northern Argentina)

These Miocene paleoprovinces and their molluscan faunas are discussed in detail in Petuch (2004: 30–43; 2012) and Petuch and Drolshagen (2009).

After a series of catastrophic events and severe climatic degenerations in the middle and late Miocene (probably caused by the Ries-Steinheim, Ewing, and Kara-Kul asteroid impacts; see Petuch, 2012; Petuch and Roberts, 2007), a new set of vibrant, species-rich paleoprovinces evolved during the early Pliocene Epoch. Being essentially extensions of the older Miocene paleoprovinces, these newly evolved biogeographical entities produced many of the characteristic endemic genera found in the Recent Carolinian, Caribbean, and Brazilian Provinces. Some of these new Pliocene groups included *Propustularia* (Cypraeidae); *Busycoarctum*, *Sinistrofugur*, and *Fulguropsis* (all Busyconidae); *Titanostrombus* (Strombidae); *Fasciolaria* (Fasciolariidae); and *Lindaconus* (Conidae). The Pliocene American paleoprovinces and paleosubprovinces include

Tropical Western Atlantic Region—*Pliocene Epoch*
    Caloosahatchian Paleoprovince (Nova Scotia to northern Mexico)
        Yorktownian Paleosubprovince (Nova Scotia to North Carolina)
        Duplinian Paleosubprovince (North Carolina to Georgia)
        Buckinghamian Paleosubprovince (southern half of Florida)
        Jacksonbluffian Paleosubprovince (northern Florida to Texas)
    Gatunian Paleoprovince (northern Mexico to southern Brazil)
        Guraban Paleosubprovince (Cuba and the West Indies)
        Veracruzan Paleosubprovince (northern Mexico to Honduras)
        Limonian Paleosubprovince (Honduras to Colombia)
        Puntagavilanian Paleosubprovince (Colombia to Suriname)
        Juruaian Paleosubprovince (Suriname to central Brazil)
    Camachoan Paleoprovince (central Brazil to central Argentina)

These Pliocene paleoprovinces and their molluscan faunas are discussed in detail in Petuch (2004: 43–53) and in Petuch and Roberts (2007). In the journal *Science*, I described a two-stage mass extinction of the Caloosahatchian molluscan faunas, at the ends of the Piacenzian and Calabrian Stages, and this may have been the result of asteroid impacts and the subsequent climatic degeneration (see Petuch, 1995b for the faunal analysis leading to this discovery).

I consider the Caloosahatchian Paleoprovince (named for the Caloosahatchee Formation of Florida) and the Gatunian Paleoprovince (named for the Gatun Formation of Panama) (see Figure 1.1) to be the direct ancestors of the Carolinian, Caribbean, Brazilian, and Panamic Provinces. Many of the modern Carolinian endemic genera, such as *Busycon* and *Sinistrofulgur* (both Busyconidae), *Cinctura* (Fasciolariidae), and *Melongena (Rexmela)* (Melongenidae) underwent large radiations within the Caloosahatchian Province. Likewise, modern Caribbean endemics such as *Muracypraea* (Cypraeidae) and Caribbean–Brazilian endemics such as *Voluta* (Volutidae) were major components of the Gatunian Province. Since the Isthmus of Panamá was open until only 3 million years ago, the Gatunian Province and its molluscan fauna spanned two oceans and extended into the eastern Pacific. After the closing of Panamá, some of these classic Gatunian groups became major components of the subsequent Carolinian, Caribbean, and Brazilian Provinces in the Atlantic and the Panamic Province in the eastern Pacific. Some of these include *Triplofusus* (Fasciolariidae; found in the Carolinian and Panamic Provinces), *Macrocypraea* (Cypraeidae; found in the Carolinian, Caribbean, Brazilian, and Panamic Provinces), and *Tenorioconus* (found in the Caribbean and Panamic Provinces). During the Pleistocene Epoch, the Caloosahatchian, Gatunian, and Camachoan molluscan faunas underwent a series of extinctions, primarily in response to severe glaciations and eustatic lows. By the mid-Pleistocene, the surviving faunal elements had established the modern provincial patterns, producing the Pleistocene Carolinian, Caribbean, and Brazilian Provinces. These distributional patterns have remained essentially the same for the past 150,000 years, resulting in the provincial and subprovincial arrangements shown throughout this book.

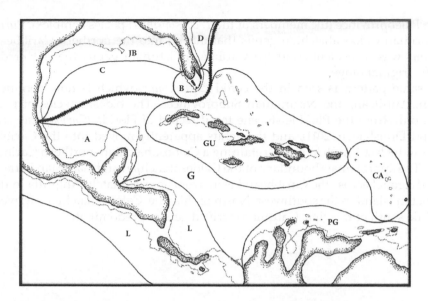

Figure 1.1 Map of the tropical western Atlantic region during the Pliocene, showing the bio-geographical arrangement prior to the closing of the Isthmus of Panama. The Pliocene coast-lines of North and South America (stippled areas) are superimposed upon the outline of the recent Americas. The paleoprovinces and paleosubprovinces include: C = Caloosahatchian Paleoprovince; D = Duplinian Paleosubprovince; B = Buckinghamian Paleosubprovince; JB = Jacksonbluffian Paleosubprovince (which encompasses the lower part of the Mississippi River Valley); G = Gatunian Paleoprovince; GU = Guraban Paleosubprovince; V = Veracruzan Paleosubprovince; L = Limonian Paleosubprovince (which spans both the Atlantic and Pacific Oceans); PG = Puntagavilanian Paleosubprovince; CA = Carriacouan Paleosubprovince. The Juruaian Paleosubprovince of the Gatunian Paleoprovince (Brazil and the Amazon River area) and the Yorktownian Paleosubprovince of the Caloosahatchian Paleoprovince (Nova Scotia to North Carolina) are not shown.

For a more detailed overview of the Miocene, Pliocene, and Pleistocene provinces and their faunas, see Petuch (2004, 2012) and Petuch and Drolshagen (2010).

One of the subtle manifestations of geographical heterochrony and paleobiogeo-graphical distributions is *paraprovincialism*, the retention of paleoprovincial and paleosub-provincial boundaries into recent time (Petuch, 1982b). These "ghosts of paleoprovincial boundaries" are present within the modern Caribbean Province and reflect the past distributions of the Pliocene Caloosahatchian and Gatunian Paleoprovinces. One of the classic examples of western Atlantic paraprovincialism is seen in the gastropod family Conidae, where the genus *Purpuriconus* is confined to the northern and western Caribbean Province (Bahamian, Antillean, and Nicaraguan Subprovinces; see Chapters 6 and 7) and where the genus *Tenorioconus* is confined to the southern provinces (Nicaraguan, Venezuelan, Grenadian, and Surinamian; see Chapters 8 and 9). The two genera over-lap in part of the Venezuelan Subprovince (the islands off Cartagena, Colombia) and in the Nicaraguan Subprovince, but there are no *Tenorioconus* species in the Bahamian and Antillean Subprovinces and no *Purpuriconus* species in the Grenadian Subprovince. This distributional pattern is also present in the Caribbean Pliocene fossil record, with *Purpuriconus* being confined to southern Florida and the Caloosahatchian Paleoprovince (*P. briani*, *P. jenniferae*, *P. erici*, and *P. protocardinalis* being the oldest known members of the genus; Petuch and Drolshagen, 2011) and with *Tenorioconus* being confined to the

Gatunian Paleoprovince (the members of the *T. consobrinus* species complex). *Purpuriconus* was a Floridian Caloosahatchian genus that moved into the northern Caribbean, while *Tenorioconus* was a resident southern Caribbean Gatunian genus that never dispersed beyond its original range.

This same pattern is seen in the conid *Cariboconus*, which is now confined to the Bahamian, Antillean, and Nicaraguan Subprovinces. The oldest species of *Cariboconus* is known only from the Plio-Pleistocene fossil record of Florida (*Cariboconus griffini*; see Petuch and Drolshagen, 2011), and the genus apparently moved into the Caribbean Sea only within the last 1 million years, retaining a Caloosahatchian–northern Caribbean distribution. The approximate boundary of the Caloosahatchian and Gatunian Paraprovinces runs at a diagonal across the Caribbean Basin, from approximately Anguilla in the northeast to Roatan Island in the southwest. North of this line, Caloosahatchian-derived genera occur, while south of this line, Gatunian-derived genera predominate.

chapter two

Provinces of the tropical western Atlantic

Of the three ecological limiting factors discussed in Chapter 1, temperature is by far the most important in the establishment of provincial boundaries. The distributions and boundaries of the three tropical western Atlantic provinces are determined almost exclusively by the surface water currents of the western halves of the *North Atlantic Gyre* and *South Atlantic Gyre* systems: with the warm *North Equatorial Current* (North Atlantic Gyre) flowing northward and dividing into the *Antilles Current, Caribbean Current*, and *Gulf Loop Current*, and later coalescing into the *Gulf Stream* (influencing the Caribbean and Carolinian Provinces); and with the warm *South Equatorial Current* (South Atlantic Gyre) flowing southward along northern South America and forming the *Brazilian Current*, which flows all along the Brazilian coast southward to Uruguay and northern Argentina (influencing the Brazilian Province). The very warm *Equatorial Countercurrent*, flowing between the North and South Equatorial Currents, carries long-lived planktotrophic larvae back and forth between West Africa and Brazil and has added a number of *amphiatlantic species* (taxa that are found on both sides of the Atlantic Ocean; discussed later in this chapter) to the western Atlantic fauna. Late Pleistocene and Holocene warm currents along the western sides of the two Atlantic gyres have created relatively stable oceanographic conditions that have allowed the evolution of the modern paratropical and eutropical provinces.

Carolinian Province

The eastern coast of the United States, from Cape Hatteras, North Carolina, south to central Florida was first considered to be a subdivision of the *Transatlantic Province* (Woodward, 1856) and was named the *Carolinian area*. This subdivision was later elevated to a more formal status, and Valentine (1973) named it the Carolinian Region. Faunal analyses of species radiations and endemic taxa, however, have shown that Woodward's and Valentine's Carolinian area actually represents a small part of a much larger biogeographical unit that extends from Cape Hatteras, around Florida and the Florida Keys, around the entire Gulf of Mexico to Isla Contoy on the Yucatan Peninsula (Figure 2.1). Observing priority of nomenclature, I here refer to the entire area as the *Carolinian Province* and have renamed the original Carolinian subdivision the *Georgian Subprovince* of the Carolinian Province (named for the State of Georgia; Petuch, 2004). In the expanded definition, four other subprovinces now compose the Carolinian Province: the Floridian, Suwannean, Texan, and Yucatanean (listed in Chapter 1 and discussed in detail in Chapters 3, 4, and 5).

The Carolinian Province, as a biogeographical unit, is bound together by an entire suite of endemic genera and families. Ecologically, the Carolinian malacofaunas are dominated by genera that originated in the Pliocene and Pleistocene Caloosahatchian Paleoprovince, and these relictual groups give the Carolinian molluscan assemblages an archaic appearance. Some of the more widespread Caloosahatchian-derived species, which are found in all five subprovinces, are shown in Figures 2.2 and 2.3 and include the following:

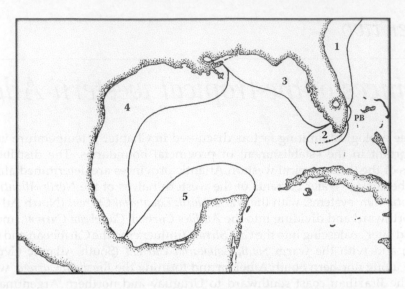

Figure 2.1 Map of the Carolinian Molluscan Province, showing the distribution of the five subprovinces: (1) Georgian Subprovince; (2) Floridian Subprovince; (3) Suwannean Subprovince; (4) Texan Subprovince; (5) Yucatanean Subprovince. The Palm Beach Provinciatone (PB), an area of faunal overlap between the Georgian and Floridian Subprovinces and the Caribbean Province, is shown along the southeastern coast of Florida.

Cypraeidae
    Macrocypraea (Lorenzicypraea) cervus
Strombidae
    Strombus alatus
Muricidae
    Hexaplex fulvescens
Fasciolariidae
    Cinctura hunteria
    Triplofusus papillosus (often listed as *Triplofusus giganteus*)
Volutidae
    Scaphella junonia (and three subspecies)
Olividae
    Oliva (Americoliva) sayana (and two subspecies)
Conidae
    Conasprelloides stimpsoni
    Dauciconus amphiurgus
    Lindaconus atlanticus
Conilithidae
    Kohniconus delessertii

Also confined to the Carolinian provincial boundaries is an entire radiation of the endemic venerid bivalve genus *Mercenaria*, including at least five species and subspecies: *Mercenaria mercenaria* (Cape Cod to northern Georgian Subprovince); *Mercenaria mercenaria notata* (Georgian Subprovince); *Mercenaria hartae* (Palm Beach Provinciatone; see Appendix 2); *Mercenaria campechiensis* (Floridian, Suwannean, Texan, and Yucatanean Subprovinces); and *Mercenaria texanum* (Texan Subprovince). Likewise, the Carolinian

Figure 2.2 Widespread Carolinian Province index species: (A) *Strombus alatus* Gmelin, 1791, length 78.5 mm. (B, C) *Macrocypraea (Lorenzicypraea) cervus* (Linnaeus, 1771), length 118 mm. (D) *Scaphella junonia* (Lamarck, 1804), length 97.6 mm. (E) *Oliva (Americoliva) sayana* Ravenel, 1834, length 77.8 mm. (F) *Cinctura hunteria* (Perry, 1811), length 48 mm.

Province houses a species radiation of the Caloosahatchian-derived *Argopecten irradians* complex (Pectinidae), including *Argopecten irradians* (Cape Cod to northern Georgian Subprovince); *Argopecten irradians concentricus* (Georgian Subprovince); *Argopecten irradians taylorae* (Floridian and Suwannean Subprovinces); and *Argopecten irradians amplicostatus* (Texan and Yucatanean Subprovinces) (the records of *Argopecten irradians amplicostatus* from northern South America represent a new, undescribed species or subspecies and not the true Carolinian *amplicostatus*). (Note: Mikkelsen and Bieler, 2007, incorrectly refer to the Florida Bay populations of *Argopecten irradians taylorae* as *Argopecten irradians concentricus*. The use of the name *taylorae* for the Florida Bay scallop is overwhelmingly supported by genetic and statistical data [Marelli et al., 1997], which Mikkelsen and Bieler apparently ignored).

The most important endemic gastropod group in the entire Carolinian Province is the family Busyconidae. Having evolved in the Caloosahatchian Paleoprovince, this large and diverse family has always been confined to eastern North America and the Gulf of Mexico and can be considered to be, collectively, the single most characteristic Carolinian molluscan taxon. Over 120 busyconid species and 12 genera are known from the fossil record (Oligocene to Pleistocene) of the eastern United States, and 17 living species and subspecies and six genera are now known from the Carolinian Province (see Petuch, 1994, 2004). For the family Busyconidae—living species and subspecies—these include the following:

Figure 2.3 Widespread Carolinian Province index species: (A) *Triplofusus papillosus* (Sowerby, 1825), length 386 mm. (B) *Sinistrofulgur sinistrum* (Hollister, 1958), length 185 mm. (C) *Kohniconus delessertii* (Recluz, 1843), length 52 mm. (D) *Lindaconus atlanticus* (Clench, 1942), length 51 mm. (E) *Dauciconus amphiurgus* (Dall, 1889), length 26.5 mm. (F) *Hexaplex fulvescens* (Sowerby, 1834), length 102 mm.

Subfamily Busyconinae
    Genus *Busycon*
        Busycon carica (Cape Cod to northern Georgian Subprovince)
        Busycon carica eliceans (southern Georgian Subprovince)
    Genus *Sinistrofulgur*
        Sinistrofulgur laeostomum (New Jersey and Georgian Subprovince)
        Sinistrofulgur sinistrum (Floridian and Suwannean Subprovinces)
        Sinistrofulgur pulleyi (Texan Subprovince)
        Sinistrofulgur perversum (Yucatanean Subprovince)
    Genus *Busycoarctum*
        Busycoarctum coarctatum (Yucatanean Subprovince)
    Genus *Lindafulgur*
        Lindafulgur candelabrum (Yucatanean Subprovince)
        Lindafulgur lyonsi (Suwannean Subprovince)
Subfamily Busycotypinae
    Genus *Busycotypus*
        Busycotypus canaliculatum (Cape Cod to Georgian Subprovince)
    Genus *Fulguropsis*
        Fulguropsis spiratum pyruloides (Georgian and Suwannean Subprovinces)
        Fulguropsis spiratum keysensis (Florida Keys, Floridian Subprovince)

Fulguropsis spiratum (Texan Subprovince)
Fulguropsis plagosus (Suwannean Subprovince)
Fulguropsis plagosus galvestonense (Texan Subprovince)
Fulguropsis plagosus texanum (Texan Subprovince)
Fulguropsis cf. *feldmanni* (Yucatanean Subprovince)

The genera *Busycon*, *Sinistrofulgur*, *Busycotypus*, and *Fulguropsis* are major predators on bivalve mollusks, often dominating sand and mud flat infaunal communities as the apex carnivores. Two of the genera, *Busycoarctum* and *Lindafulgur*, are primary predators in deep-water sand-bottom communities in the Gulf of Mexico. Another important Carolinian endemic radiation is seen in the genus *Cinctura*, the banded tulip shells (Fasciolariidae), which contains at least five species and subspecies and is confined to the provincial boundaries: *Cinctura hunteria* (Georgian, Floridian, Suwannean, and Texan Subprovinces); *Cinctura hunteria keatonorum* (offshore subspecies; Georgian Subprovince); *Cinctura tortugana* (Floridian Subprovince); *Cinctura lilium* (Texan Subprovince); and *Cinctura branhamue* (Yucatanean Subprovince).

Faunal analysis of Carolinian mollusks

As outlined in Chapter 1, the Carolinian Province, as a discrete biogeographical entity, can be defined by the taxon index for each of the 10 key provincial index taxa. The raw data for these analyses are taken from the species list shown in Appendix 1. Based upon the relationship

$$T = \frac{n}{N}(100),\ T > 50$$

these are as follows:

Modulidae (*T1*), $T = 100$, with $N = 5$, $n = 5$
Turbinellidae (*T2*), $T = 50$, with $N = 4$, $n = 2$
Conidae (*T3*), $T = 66$, with $N = 39$, n = 29
Conilithidae (*T4*), $T = 92$, with $N = 12$, $n = 11$
Muricinae (*T5*), $T = 58$, with $N = 31$, $n = 18$
Fasciolariinae (*T6*), $T = 87$, with $N = 8$, $n = 7$
Lyriinae (*T7*), $T = 100$, with $N = 1$, $n = 1$
Olivinae (*T8*), $T = 87$, with $N = 8$, $n = 7$
Cancellariinae (T9), $T = 55$, with $N = 9$, $n = 5$
Plesiotritoninae (*T10*), $T = 50$, with $N = 2$, $n = 1$

The provincial combined index, incorporating the endemicity of these 10 families and subfamilies and as defined by the relationship

$$P = \sum_{n=1}^{10} \frac{Tn}{10},\ P > 50$$

yields the quantity $P = 74.5$. This high provincial index (with $P > 50$) demonstrates that the Carolinian, with its five subprovinces, is a strong province exhibiting a high level of endemism. Based on the taxonomic and distributional data shown here, Woodward's Carolinian subdivision of the Transatlantic Province (1856) and Valentine's Carolinian Province and

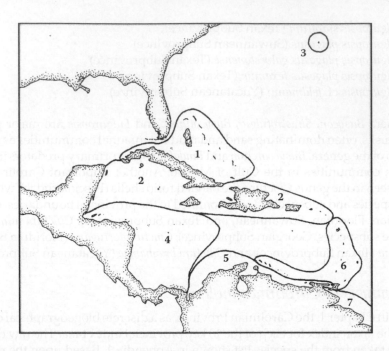

Figure 2.4 Map of the Caribbean Molluscan Province, showing the distribution of the seven sub-provinces: (1) Bahamian Subprovince; (2) Antillean Subprovince; (3) Bermudan Subprovince; (4) Nicaraguan Subprovince; (5) Venezuelan Subprovince; (6) Grenadian Subprovince; (7) Surinamian Subprovince.

Gulf Province (1973) are not substantiated and should be considered only sections of the Carolinian Province. The endemic species radiations of the Carolinian Province will be discussed in detail in Chapters 3, 4, and 5.

Caribbean Province

The area extending from Bermuda to the Bahamas and Cuba and extending across the Antilles Island Arc and the Caribbean Sea Basin south to Suriname and the mouth of the Amazon River is here considered to compose the Caribbean Province (Figure 2.4). As a biogeographical entity, the Caribbean Province was recognized by Woodward (1856), Warmke and Abbott (1961), Valentine (1973), and Briggs (1995), but in either expanded or truncated forms that do not reflect the actual spatial arrangement of molluscan faunas. As in the case of the Carolinian Province, the Caribbean Province is bound together by an entire suite of endemic genera, families, and species complexes. Ecologically, Caribbean malacofaunas are dominated by genera that originated in the Pliocene Gatunian Paleoprovince, and many of these have persisted into the recent in several secondary relict pockets that occur within the provincial boundaries. Some of these widespread Gatunian-derived taxa, which are found in all or most of the subprovinces of the Caribbean Province, include (some of which are shown in Figure 2.5):

Family Ovulidae
    Cyphoma gibbosum
    Cyphoma signatum
    Cyphoma intermedium

Figure 2.5 Widespread Caribbean Province index species: (A) *Dauciconus daucus* (Hwass, 1792), length 28 mm. (B) *Strombus pugilis* Linnaeus, 1758, length 82.6 mm. (C) *Eustrombus gigas* (Linnaeus, 1758), length 240 mm. (D) *Gladioconus mus* (Hwass, 1792), length 29 mm. (E) *Cyphoma gibbosum* (Linnaeus, 1758), length 28.7 mm. (F) *Oliva (Americoliva) reticularis* (Lamarck, 1810), length 36 mm.

Family Strombidae
   Strombus pugilis
   Eustrombus gigas
Family Cassidae
   Cassis flammea
   Cassis madagascariensis
Family Muricidae
   Chicoreus brevifrons
   Phyllonotus pomum
Family Olividae
   Oliva (Americoliva) reticularis
   Oliva (Americoliva) bifasciata
Family Conidae
   Dauciconus daucus
   Gladioconus mus

Unlike the Carolinian Province, the Caribbean exhibits much more regional endemism, with many genera and species complexes being confined to only one subprovince. Particularly noteworthy are the *Purpuriconus* (Conidae) species radiation of the Bahamian Subprovince, with over 14 species found on the Bahama Banks, Turks and Caicos, and northern Cuba; the *Gradiconus* (Conidae) species radiation of the Nicaraguan Subprovince, with at least seven species being found along Nicaragua and Panamá; and the *Tenorioconus* (Conidae) species radiation of the Grenadian Subprovince, with at least 10 species and sub-species being found on the islands along northern South America and the Lesser Antilles. The endemic species radiations of the Caribbean Province will be discussed in detail in Chapters 6, 7, 8, and 9.

Faunal analyses of Caribbean mollusks

The Caribbean Province, as a discrete biogeographical entity, can be defined by the taxon index of each of the 10 key provincial index taxa. The raw data used for these analyses are taken from the species lists shown in Appendix 1. Based upon the relationship

$$T = \frac{n}{N}(100), \, T > 50$$

these are as follows:

　　Modulidae (*T1*), $T = 75$, with $N = 4$, $n = 3$
　　Turbinellidae (*T2*), $T = 80$, with $N = 5$, $n = 4$
　　Conidae (*T3*), $T = 92$, with $N = 111$, $n = 102$
　　Conilithidae (*T4*), $T = 91$, with $N = 23$, $n = 21$
　　Muricinae (*T5*), $T = 76$, with $N = 70$, $n = 55$
　　Fasciolariinae (*T6*), $T = 66$, with $N = 3$, $n = 2$
　　Lyriinae (*T7*), $T = 100$, with $N = 11$, $n = 11$
　　Olivinae (*T8*), $T = 95$, with $N = 21$, $n = 20$
　　Cancellariinae (*T9*), $T = 50$, with $N = 8$, $n = 4$
　　Plesiotritoninae (*T10*), $T = 66$, with $N = 3$, $n = 2$

　　The provincial combined index, incorporating the endemicity of these 10 families and subfamilies and defined by the relationship

$$P = \sum_{n=1}^{10} \frac{Tn}{10}, \, P > 50$$

yields the index quantity $P = 86.5$. Based on this large value (with $P > 50$), the Caribbean Province, as defined here, is a strong biogeographical entity that exhibits a high level of endemicity.

Brazilian Province

Until 1988, when I formally proposed and described the Brazilian Province, this large and oceanographically complex area was considered to be a southern extension of the Caribbean Province (Warmke and Abbott, 1961) or a split area, with the northern half belonging to the Caribbean Province and the southern half belonging to the Patagonian Province (Woodward, 1856; Valentine, 1973; Briggs, 1995). The Brazilian Province (Figure 2.6), which extends from the mouth of the Amazon River southward to Mar del Plata, Argentina, is oceanographically and ecologically separated from the Caribbean Province by the barrier of

Figure 2.6 Map of the Brazilian Molluscan Faunal Province, showing its three subprovinces and southern provinciatone: (1) Cearaian Subprovince; (2) Bahian Subprovince; (3) Paulinian Subprovince; (4) Uruguayan Provinciatone. Areas of special biogeographical interest include the following: AM = the mouth of the Amazon River; R = Atol das Rocas Archipelago; N = Fernando de Noronha Islands; S = Salvador, Bahia State, and Todos os Santos Bay; A = Abrolhos Archipelago and reef complexes; T = Trindade Island; RJ = Rio de Janeiro; F = Florianopolis, Santa Catarina State.

the copious fresh water effluent emanating from the Amazon River. The Brazilian malaco-fauna derives from the Juruaian Paleosubprovince of the Pliocene Gatunian Paleoprovince and contains a distinctive fauna that is more similar to that of the Eastern Pacific Panamic Province than it is to the Caribbean Province fauna. Such typical Panamic genera as the gastropod *Caducifer* (Buccinidae) and the bivalve *Miltha* (Lucinidae) occur in the Brazilian Province but are not present in the Caribbean Province. The Brazilian cowrie *Macrocypraea (Macrocypraea) zebra dissimilis*, although considered a subspecies of the Caribbean *M. (M.) zebra*, most often lacks the well-formed color rings on the base of the shell and more closely resembles, in shape and color pattern, the Panamic *Macrocypraea (Macrocypraea) cervinetta*. Likewise, the strongly callused and knobbed Brazilian flamingo tongue, *Cyphoma macumba*, more closely resembles the Panamic *Cyphoma emarginatum* than it does any of the congeneric Caribbean species. The Brazilian Province also contains a number of distinctive endemic gastropod genera, some of which include the volutid *Plicoliva*, the conid

Figure 2.7 Widespread Brazilian Province index species: (A) *Vasum cassiforme* (Kiener, 1841), length 89 mm. (B) *Strombus worki* Petuch, 1993, length 81 mm. (C) *Siratus tenuivaricosa* Dautzenberg, 1927, length 61 mm. (D) *Pleuroploca aurantiaca* (Lamarck, 1816), length 86 mm. (E) *Siratus senegalensis* (Gmelin, 1791), length 55 mm. (F) *Thaisella mariae* (Morretes, 1954), length 26 mm.

Lamniconus, and the conilithids *Artemidiconus* and *Coltroconus.* Some of the widespread endemic taxa that are found in all or most of the subprovinces of the Brazilian Province include (some of which are shown in Figure 2.7) the following:

Strombidae
  Strombus worki
Cypraeidae
  Luria cinerea brasiliana
  Macrocypraea (Macrocypraea) zebra dissimilis
Bursidae
  Bursa ponderosa
Muricidae
  Siratus senegalensis
  Siratus tenuivaricosa
  Thaisella mariae
Fasciolariidae
  Leucozonia brasiliana
  Pleuroploca aurantiaca
Buccinidae
  Pisania janeirensis

Turbinellidae
  Turbinella laevigata
  Vasum cassiforme
Olividae
  Oliva (Americoliva) circinata (with two subspecies)

Of special interest in the southern subprovinces of the Brazilian Province is a large radiation of the endemic cone shell genus *Lamniconus* (Conidae), including at least six species: *Lamniconus clerii* (Paulinian Subprovince); *Lamniconus clenchi* (Paulinian Subprovince); *Lamniconus lemniscatus* (Paulinian Subprovince); *Lamniconus xanthocinctus* (southern Bahian and northern Paulinian Subprovinces); *Lamniconus tostesi* (Paulinian Subprovince); and *Lamniconus carcellesi* (southern Paulinian Subprovince and Uruguayan Provinciatone). Likewise, the Brazilian Province houses an endemic species radiation of the Caribbean–Brazilian cone shell genus *Poremskiconus*, including at least seven species having been described: *Poremskiconus mauricioi* (Cearaian Subprovince); *P. archetypus* (southern Bahian and northern Paulinian Subprovinces); *P. bertarollae* (southern Bahian Subprovince); *P. abrolhosensis* (= *baiano*) (southern Bahian Subprovince); *P. brasiliensis* (southernmost Bahian Subprovince); and *P. cargilei* (Bahian Subprovince). (Note: For the description of the new genus *Poremskiconus*, see Appendix 2.) The southern subprovinces of the Brazilian Province also harbor large endemic species radiations of the volutid genus *Odontocymbiola*, with at least seven species, and the olivid genus *Olivancillaria*, with at least nine species. These, and other endemic genera and species radiations of the Brazilian Province, will be discussed in detail in Chapters 10 and 11.

Faunal analyses of Brazilian mollusks

The Brazilian Province, as a discrete biogeographical entity, can be defined by the taxon index of each of the 10 key provincial index taxa. The raw data used for these analyses are taken from the species lists shown in Appendix 1. Based on the relationship

$$T = \frac{n}{N}(100),\ T > 50$$

these are as follows:

Modulidae (*T1*), $T = 50$, with $N = 2$, $n = 1$
Turbinellidae (*T2*), $T = 100$, with $N = 2$, $n = 2$
Conidae (*T3*), $T = 87$, with $N = 23$, $n = 20$
Conilithidae (*T4*), $T = 82$, with $N = 11$, $n = 9$
Muricinae (*T5*), $T = 60$, with $N = 18$, $n = 10$
Fasciolariinae (*T6*), $T = 75$, with $N = 4$, $n = 3$
Lyriinae (*T7*), $T = 100$, with $N = 1$, $n = 1$
Olivinae (*T8*), $T = 100$, with $N = 3$, $n = 3$
Cancellariinae (*T9*), $T = 100$, with $N = 3$, $n = 3$
Plesiotritoninae (*T10*), $T = 50$, with $N = 2$, $n = 1$

The provincial combined index, incorporating the endemicity of these 10 families and subfamilies and defined by the relationship

$$P = \sum_{n=1}^{10} \frac{Tn}{10}, \ P > 50$$

yields the index quantity $P = 82$. As in the Carolinian and Caribbean Provinces, this high level of endemism (with $P > 50$) supports the full provincial status of the Brazilian Province. The very large index quantity also demonstrates that the Brazilian Province is a strong biogeographical entity and is not simply a southern component of the Caribbean Province. This paratropical province, as defined here, is a fusion of Woodward, Valentine, and Briggs's old Caribbean Province (southern part) and Patagonian Province. The true Patagonian Province, as recognized here, extends south of the Brazilian Province and encompasses the entire coast of Patagonian Argentina, from Mar del Plata south to the Golfo San Jorge. The subantarctic Magellanic Province extends south of the Patagonian, from the Golfo San Jorge to the Falkland Islands and around the Straits of Magellan to Chiloe Island, Chile. See Rios (1994) for an illustrated checklist of Brazilian shells (with much of the taxonomy, however, being outdated).

Western Atlantic amphiprovincial mollusks

A large number of marine gastropods and bivalves have extremely wide ranges across the tropical western Atlantic, with some occurring in the Carolinian, Caribbean, and Brazilian Provinces. All of these wide-ranging taxa have long-lived planktotrophic larvae (being part of the meroplankton) that can travel great distances along the American coastlines before settling out and metamorphosing into adults. These widespread species, whose ranges span all three tropical provinces, are here referred to as *amphiprovincial taxa*. One of the most prominent amphiprovincial groups is the Cypraeidae, the cowrie shells. Having extremely long-lived meroplanktonic sinusigera veligers, the cowries have the ability to ride the currents for months and disperse themselves all across the tropical western Atlantic, from the Georgian Subprovince in the north to the Paulinian Subprovince in the south. One species in particular, *Erosaria acicularis* (see Figure 2.8), covers virtually the entire tropical western Atlantic, ranging from Cape Hatteras, North Carolina, and Bermuda south to Santa Catarina, Brazil, and all the Brazilian offshore islands. Some of the more important amphiprovincial mollusks, being found in all three provinces, are listed here (with several illustrated in Figures 2.8, 2.9, and 2.10):

Cerithiidae
    Cerithium atratum (also amphiatlantic, being found in West Africa)
    Cerithium eburneum
    Cerithium guinaicum (also amphiatlantic, being found in West Africa)
    Cerithium litteratum
Cypraeidae
    Erosaria acicularis
    Luria cinerea (Carolinian, Caribbean, northern Brazilian)
    Macrocypraea (Macrocypraea) zebra (Carolinian, Caribbean, northernmost Brazilian)
    Propustularia surinamensis (southern Carolinian, Caribbean, Brazilian)
Tonnidae
    Tonna galea (Carolinian and Caribbean; also amphiatlantic, being found in West
        Africa and the Mediterranean Sea; with a subspecies, *T. galea brasiliana* in Brazil)

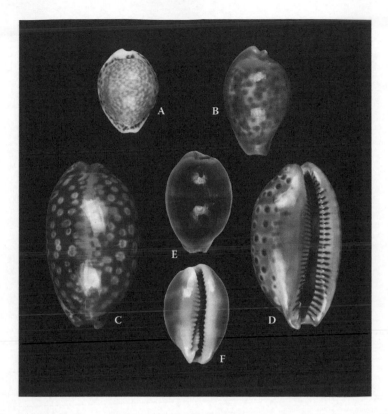

Figure 2.8 Western Atlantic amphiprovincial Cypraeidae: (A) *Erosaria acicularis* (Gmelin, 1791), length 24 mm. (B) *Propustularia surinamensis* (Perry, 1811), length 26.8 mm. (C, D) *Macrocypraea (Macrocypraea) zebra* (Linnaeus, 1758), length 61.3 mm. (E, F) *Luria cinerea* (Gmelin, 1791), length 24 mm.

Cassidae
  Cassis tuberosa (also Cape Verde Islands)
  Phalium granulatum
Muricidae
  Phyllonotus oculatus
Fasciolariidae
  Fasciolaria tulipa (Carolinian, Caribbean, and northern Brazilian)
Conidae
  Chelyconus ermineus (also amphiatlantic, being found in West Africa)
  Stephanoconus regius
Bullidae
  Bulla striata (also amphiatlantic, being found in West Africa)
Bivalvia
  Lindapecten muscosus

(Note: Mikkelsen and Bieler, 2007, incorrectly placed *Lindapecten muscosus* in the genus *Aequipecten* and ignored the genus *Lindapecten*. Waller, 2011, showed that *Lindapecten* is a valid taxon and should be used as the genus for *muscosus*.)

Figure 2.9 Western Atlantic amphiprovincial gastropods: (A) *Cerithium atratum* Born, 1778, length 32 mm. (B) *Chelyconus ermineus* (Born, 1778), yellow color form, length 40 mm. (C) *Cassis tuberosa* (Linnaeus, 1758), length 173 mm. (D) *Stephanoconus regius* (Gmelin, 1791), length 50 mm. (E) *Bulla striata* Bruguiere, 1792, length 24 mm. (F) *Fasciolaria tulipa* (Linnaeus, 1758), length 180 mm.

As provincial and subprovincial indicators, these amphiprovincial species are essentially useless and can be considered "background noise" in faunal surveys and analyses. Because of their broad distributions and ecological plasticity, however, amphiprovincial taxa such as these are useful in defining the extreme high latitude limits of the subtropical areas of the western Atlantic.

Figure 2.10 Close-up of an aggregation of the cerithiid gastropod *Cerithium litteratum* (Born, 1778) in a tide pool on the rocky shoreline along Missouri Key, Florida Keys. This small and abundant shallow water gastropod is a classic amphiprovincial mollusk, being found in the Carolinian, Caribbean, and Brazilian Provinces. (From Petuch, E.J., and D.M. Sargent, *Visaya*, 3(4), 98–104, 2011a.)

chapter three

Molluscan biodiversity in the Georgian Subprovince

Extending from Cape Hatteras, North Carolina, south to Palm Beach County, Florida, the Georgian Subprovince encompasses the area that Woodward (1856) referred to as the Carolinian subdivision of the Transatlantic Province and that Valentine (1973) called the Carolinian Province. Although it contains all of the classic provincial index taxa listed in Chapter 2, this eastern U.S. faunal area also exhibits a high level of endemism and can be differentiated from the other Carolinian subprovinces. As a discrete biogeographical entity, the Georgian Subprovince (including the Palm Beach Provinciatone) can be defined by the percentages of endemism of the 10 key index taxa. Based on the relationship

$$T = \frac{n}{N}(100), \ T > 50$$

these are as follows:

Modulidae ($t1$), $t = 100$, with $N = 4$, $n = 4$
Turbinellidae ($t2$), $t = 0$, with $N = 1$, $n = 0$ (absent in the Georgian Subprovince)
Conidae ($t3$), $t = 29$, with $N = 17$, $n = 5$
Conilithidae ($t4$), $t = 17$, with $N = 6$, $n = 1$
Muricinae ($t5$), $t = 21$, with $N = 14$, $n = 3$
Fasciolariinae ($t6$), $t = 33$, with $N = 3$, $n = 1$
Lyriinae ($t7$), $t = 0$, with $N = 0$, $n = 0$ (absent in the Georgian Subprovince)
Olivinae ($t8$), $t = 33$, with $N = 3$, $n = 1$
Cancellariinae ($t9$), $t = 0$, with $N = 3$, $n = 0$
Plesiotritoninae ($t10$), $t = 50$, with $N = 2$, $n = 1$

The raw data for these analyses are taken from the species lists shown in Appendix 1. The subprovincial combined index, incorporating the endemicity of these 10 families and subfamilies and defined by the relationship

$$S = \sum_{n=1}^{10} \frac{tn}{10}, \ S > 25$$

yields the index quantity $S = 30.2$. This high level of subprovincial endemism (with $S > 25$) demonstrates that the Georgian Subprovince is a differentiable biogeographic entity.

The Georgian Subprovince contains a number of characteristic habitats and environments, all of which support distinctive ecosystems and molluscan assemblages. North of Florida, the entire shoreline of the subprovincial area is dominated by coastal lagoons behind barrier island systems. Because of strong seasonal temperature and salinity fluctuations, these shallow lagoonal environments harbor relatively low-diversity malacofaunas, but ones that still exhibit high levels of endemism. Much more species-rich communities lie just offshore of the lagoonal habitats, where northward-flowing warm currents stabilize the marine environmental parameters. Two of the most prominent types of communities found in these stable offshore environments are one based on immense aggregations of scallops and another associated with coral bioherms growing along the edge of the continental shelf. The malacofaunas associated with these Georgian lagoonal and coral bioherm environments become richer and ecologically more diverse farther south within the Palm Beach Provinciatone transition zone. The subprovincial combined index is a compilation of species from all of these habitats.

Carolinas and Georgia coastal lagoons

From Cape Hatteras south to the Georgia–Florida border, the physiography of the Georgian Subprovince is dominated by large coastal lagoons behind extensive barrier island systems. Fed by braided streams flowing down from the Appalachian Mountains, these lagoons receive a large input of muddy sediments and fresh water during the spring. In the summer months, these enclosed lagoons receive less freshwater effluent and often develop hypersaline conditions and high water temperatures. In the winter, lagoonal salinities remain relatively constant, but water temperatures often plummet to below 20 degrees Celsius, producing temperate water conditions.

These seasonal fluctuations have led to the evolution of physiologically plastic organisms, which are able to tolerate wide ranges of salinity and temperature (euryhaline and eurythermal mollusks). Because of these seasonal nontropical conditions, several species from higher latitudes have managed to establish themselves within the coastal environments along the northernmost edge of the Carolinian Province. Most of these northern-derived species are migrants from the Virginian Province, which extends from Cape Cod, Massachusetts, south to Cape Hatteras.

The following is a list of some of the more prominent and characteristic species and subspecies found within the coastal lagoons of the Georgian Subprovince (several illustrated in Figure 3.1):

Gastropoda
Crepidulidae
   Crepidula fornicata (also in the Virginian Province)
Naticidae
   Neverita duplicata (also in the Virginian Province)
   Sinum perspectivum
Muricidae
   Eupleura caudata (also in the Virginian Province)
   Stramonita floridana
   Urosalpinx cinerea (also in the Virginian Province)
Fasciolariidae
   Cinctura hunteria
Busyconidae
   Busycon carica (also in the Virginian Province)
   Busycon carica eliceans
   Busycotypus canaliculatum (also in the Virginian Province)
   Fulguropsis spiratum pyruloides
   Sinistrofulgur laeostomum (also found in the southern part of the Virginian Province)

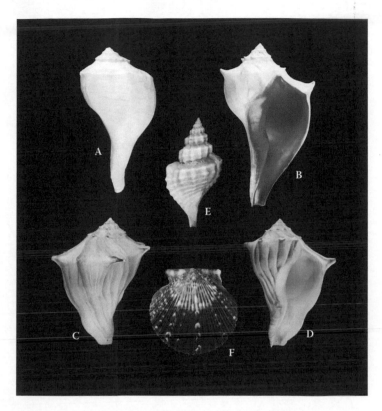

Figure 3.1 Mollusks of the Georgian Subprovince coastal lagoons: (A) *Sinistrofulgur laeostomum* (Kent, 1982), length 190 mm. (B) *Busycon carica* (Gmelin, 1791), length 230 mm. (C, D) *Busycon carica eliceans* (Montfort, 1810), length 135 mm. (E) *Eupleura caudata* (Say, 1822), length 28 mm. (F) *Argopecten irradians concentricus* (Say, 1822), length 51 mm.

Terebridae
  Strioterebrum concava
  Strioterebrum dislocata
  Strioterebrum cf. protexta
Bivalvia
Pectinidae
  Argopecten irradians concentricus
Ostreidae
  Crassostrea virginica (also in the Virginian Province)
Veneridae
  Mercenaria mercenaria notata

The only macromollusk that is endemic to the Georgian Subprovince coastal lagoon systems is the busyconid *Busycon carica eliceans* (Figure 3.1C,D). This predatory gastropod is generally considered to be a subspecies of the Virginian Province *Busycon carica* and differs from the nominate subspecies in having a smaller, broader, and heavier shell and in having a much wider, more flaring siphonal canal. While *Busycon carica carica* ranges from Cape Cod to northern North Carolina and offshore all the way to northeastern Florida, *Busycon carica eliceans* is confined to the coastal lagoons of South Carolina and Georgia.

Figure 3.2 Mollusks of the Georgian Subprovince offshore scallop beds: (A, B) *Vokesimurex morrisoni* Petuch and Sargent, 2011, length 39.5 mm. (C, D) *Gradiconus philippii* (Kiener, 1845), length 30 mm. (E) *Argopecten gibbus carolinensis* Grau, 1952, length 49 mm. (F) *Polygona williamlyonsi* Petuch and Sargent, 2011, length 45.8 mm.

Carolinas and Georgia offshore scallop beds

The middle neritic zone (20–100 m depths) of the Georgian Subprovince, from Cape Hatteras to east central Florida, houses one of the richest benthonic communities found anywhere along the eastern United States. Unique to the Georgian Subprovince, this distinctive ecosystem and its associated molluscan assemblage is centered on *Argopecten gibbus carolinensis*, a large endemic subspecies of the southern Carolinian–Caribbean calico scallop (Figure 3.2E). Forming immense shoals just offshore, the Carolina calico scallop is the basis of a huge commercial scallop fishing industry and is one of the principal seafood exports of North Carolina. The wide, flat continental shelf, warm Gulf Stream waters, and the rich effluent of nutrients draining out of the coastal lagoons and braided streams all combine to create the perfect environmental conditions to support such an immense molluscan biomass.

These offshore scallop beds contain a full complement of Carolinian index taxa, along with several endemic scallop-associated species. Some of the more prominent and important taxa are listed here (several shown in Figure 3.2):

Gastropoda
Strombidae
 Strombus alatus

Cassidae
  Cassis spinella
  Phalium granulatum
Personiidae
  Distorsio clathrata
Ranellidae
  Cymatium parthenopeum
Muricidae
  Hexaplex fulvescens
  Vokesimurex morrisoni (Georgian Subprovince endemic)
Fasciolariidae
  Cinctura hunteria keatonorum (see Appendix 2)
  Polygona williamlyonsi (Georgian endemic, also found in the Palm Beach Provinciatone)
  Triplofusus papillosus
Busyconidae
  Fulguropsis spiratum pyruloides
Volutidae
  Scaphella junonia
Olividae
  Oliva (Americoliva) bifasciata hollingi
  Oliva (Americoliva) sayana
Conidae
  Conasprelloides stimpsoni
  Dauciconus amphiurgus
  Gradiconus philippii (Georgian Subprovince endemic)
  Lindaconus atlanticus
Conilithidae
  Jaspidiconus pfluegeri (Georgian Subprovince endemic; also Palm Beach Provinciatone)
  Kohniconus delessertii
Bivalvia
Pectinidae
  Argopecten gibbus carolinensis (Georgian Subprovince endemic)
  Nodipecten fragosus
  Pecten ziczac
Cardiidae
  Laevicardium laevigatum

Carolinas and Georgia offshore coral bioherms

In 1969, Ian Macintyre of the Smithsonian Institution and Orrin Pilkey of Duke University first published an amazing discovery: the farthest north coral reef complexes in the Atlantic Ocean, found growing along the edge of the continental shelf off North Carolina. Occurring as far north as Cape Hatteras and Cape Lookout, these coral bioherms (non-zonated patch reefs) were found to be composed of only two types of hermatypic scler-actinians: the star corals *Siderastrea siderea* and *Solenastrea hyades*. These two reef-building coral species grew together with the ahermatypic scleractinian *Oculina diffusa* to produce a high-latitude, true Caribbean-type reef system. Growing on shallow (10–15 m depths) banks and ridges in the outermost neritic zone, these coral bioherms are constantly bathed by the warm waters of the Gulf Stream and exist in subtropical oceanographic conditions.

Figure 3.3 Gastropods of the Georgian Subprovince offshore coral bioherms: (A) *Gemophos filis-triatus* Vermeij, 2006, length 20.4 mm. (B) *Pusula juyingae* Petuch and Sargent, 2011, length 16.7 mm. (C) *Modulus lindae* Petuch, 1987, height 11 mm. (D) *Modulus kaicherae* Petuch, 1987, width 12 mm. (E) *Favartia goldbergi* Petuch and Sargent, 2011, length 10.5 mm.

Subsequent research showed that the reef systems were not unique to North Carolina but extended southward along the edge of the continental shelf as far as Florida. Because the continental shelf off the Carolinas and Georgia is so wide, these northern reefs are as much as 70 miles (120 km) offshore and are rarely, if ever, visited by researchers or divers.

I was the first to publish about the molluscan fauna associated with these offshore coral reefs (Petuch, 1972, 1974a), based on a compilation of data collected in 1971 during my cruise on the Duke University research vessel *R/V Eastward*. A completely unsuspected complement of tropical gastropods was collected on these reefs, including the large milk conch *Macrostrombus costatus* (Strombidae, which was found to be very common around the reefs) and the rare Morum *Cancellomorum dennisoni* (Harpidae-Moruminae, previously known only from the Caribbean region). The following is a list of some of the more prominent and important Georgian Subprovince reef-associated gastropods (several shown in Figure 3.3):

Modulidae
    Modulus kaicherae (also Palm Beach Provinciatone reefs)
    Modulus lindae (also Palm Beach Provinciatone reefs)
Cypraeidae
    Erosaria acicularis
    Luria cinerea

    Macrocypraea (Lorenzicypraea) cervus
    Macrocypraea (Macrocypraea) zebra
Ovulidae
    Cymbovula acicularis
    Cyphoma gibbosum
    Cyphoma macgintyi
Triviidae
    Pusula juyingae (Palm Beach Provinciatone reefs, possibly off Georgia)
Cassidae
    Cassis spinella
    Phalium granulatum
Strombidae
    Macrostrombus costatus
Muricidae
    Favartia goldbergi (also Palm Beach Provinciatone reefs)
    Hexaplex fulvescens
    Murexiella leviculus
    Pazinotus stimpsoni
    Phyllonotus pomum
    Vokesimurex rubidus
Buccinidae
    Gemophos filistriatus (also Palm Beach Provinciatone reefs)
Harpidae
    Cancellomorum dennisoni
Marginellidae
    Prunum cineraceum
Conidae
    Chelyconus ermineus
    Gladioconus mus
    Stephanoconus regius
Conilithidae
    Jaspidiconus mindanus
Terebridae
    Strioterebrum onslowensis
Mangelliidae
    Cryoturris elata
Drilliidae
    Drillia wolfei

Since the Georgian Subprovince reefs are almost completely unexplored and unstudied, future research on these northernmost coral complexes will doubtlessly yield many new and fascinating discoveries.

Georgian deep-water areas

The upper bathyal zone off the Georgian Subprovince contains a rich, yet virtually unexplored, molluscan fauna. What little has been published on this area, mostly from fisheries research surveys conducted in the late 1890s and early 1900s, has shown that a high level of endemicity is present, with many examples of submergence of previously shallow water taxa.

Figure 3.4 Deep-water Volutoidean gastropods of the Georgian Subprovince: (A, B) *Aurinia (Rehderia) georgiana* (Clench, 1946), length 80 mm. (C, D) *Scaphella (Clenchina) gouldiana* (Dall, 1887), length 54 mm. (E) *Prunum canillum* (Dall, 1927), length 16 mm.

Some of these include the endemic marginellid *Prunum canillum* and the endemic volutid *Aurinia (Rehderia) georgiana*, both of which are descendants of groups that lived in shallow water during the Miocene and Pliocene. In some shallower areas of the Georgian upper bathyal zone, just below the neritic–bathyal zone break at around 250–300 m depths, the ahermatypic oculinid scleractinian *Lophohelia* forms extensive coral biohermal structures. These "deep-water coral reefs" house an interesting molluscan fauna that is unique within the Carolinian Province. Since very little research has been conducted on the upper bathyal zone and *Lophohelia*-associated molluscan faunas, only sketchy and incomplete faunal lists are available for these areas. Of special interest in the bathyal areas of the Georgian Subprovince is a large endemic species radiation of the daphnellid turroidean genera *Pleurotomella* and *Majox* (Petuch, 1974b), which have evolved at least 15 species. Most of these large-shelled turroideans are clustered along either the areas off the Carolinas or the areas off Georgia and northeastern Florida. The following is a list of some of the more prominent and important Georgian upper bathyal zone gastropods (with some typical volutoidean shown on Figure 3.4):

Pleurotomariidae *Perotrochus maureri*
Calliostomatidae
    Calliostoma bairdi
    Calliostoma benedicti
    Calliostoma psyche
    Calliostoma sayanum

Cassidae
  Echinophoria coronadoi
Tonnidae
  Oocorys abyssorum
  Oocorys sulcata
Columbariidae
  Fulgurofusus timor (endemic to the Carolinas; off Cape Fear)
Muricidae
  Murexiella hidalgo
  Pterynotus phaneus
  Pteropurpura bequaerti
  Urosalpinx macra (possibly a living member of the fossil genus Trossulasalpinx)
Fasciolariidae
  Fusinus schrammi
  Harasewychia vitreus
Buccinidae
  Buccinum abyssorum
  Liomesus stimpsoni
  Mohnia carolinensis
Volutomitridae
  Latiromitra bairdii (endemic to the Carolinas)
Volutidae
  Aurinia (Rehderia) georgiana (endemic; also found in the Palm Beach Provinciatone)
  Bathyaurinia aguayoi (endemic to the southern part of the Georgian Subprovince)
  Scaphella (Clenchina) gouldiana (endemic to the Georgian Subprovince; this species was erroneously reported from Texas by Tunnell et al., 2010: 226)
Marginellidae
  Hyalina styria (endemic to Georgia and northeastern Florida)
  Prunum canillum (endemic to Georgia and northeastern Florida)
Cancellariidae
  Axelella agassizii
  Axelella smithi (endemic to the Carolinas)
Conilithidae
  Dalliconus macgintyi
Turridae
  Gemmula periscelida
  Leucosyrinx tenoceras
  Leucosyrinx verrillii
Drilliidae
  Inodrillia hattersensis (endemic to the Carolinas)
Daphnelliidae
  Majox arestum (endemic to Georgia and northeastern Florida)
  Majox bairdi costlowi (endemic to the Carolinas; incorrectly referred to as *Bathybela tenelluna* [Locard, 1897], which is an eastern Atlantic and Mediterranean species)
  Majox chariessum (endemic to the Carolinas)
  Majox phalerum (endemic to the Carolinas)
  Pleurotomella aperta (endemic to Georgia and northeastern Florida)
  Pleurotomella atypha (endemic to the Carolinas)
  Pleurotomella corrida (endemic to Georgia and northeastern Florida)

Pleurotomella dalli (endemic to the Carolinas)
Pleurotomella hadria (endemic to the Carolinas)
Pleurotomella leptalea (endemic to the Carolinas)
Pleurotomella lineola (endemic to Georgia northeastern Florida)
Pleurotomella stearina (endemic to Georgia and northeastern Florida)
Pleurotomella sulcifera (endemic to the Carolinas)
Pleurotomella vaginata (endemic to Georgia and northeastern Florida)

Palm Beach Provinciatone

At the extreme southern end of the Georgian Subprovince, in the area of Florida extending from approximately Fort Pierce, St. Lucie County, south to northern Broward County (Petuch, 1980b), a broad zone of faunal overlap exists between the two eastern subprovinces of the Carolinian Province. Along this stretch of coastline, the Gulf Stream comes very close to the shore, with its closest approach being in Palm Beach County, from approximately Lake Worth northward to Jupiter. Because of the proximity of this warm current, the area is exposed to tropical water conditions for most of the year but, because of its high latitude, is frequently exposed to cold temperatures for brief times during the winter. These extremes in temperature, particularly for shallow-water or intertidal mollusks, limit the types of species that can inhabit this area. As in the case of all provinciatones, endemic species evolve that can tolerate and thrive in the fluctuating climates, and these often dominate the local molluscan assemblages. The cool winter temperatures also allow physiologically plastic high latitude species to migrate into the provinciatonal area, where they live sympatrically with low latitude tropical taxa.

Two main environments predominate in the area of the Palm Beach Provinciatone: large coastal lagoons behind barrier islands and deep offshore reefs that are growing on submerged Pleistocene beach rock terraces (see Petuch, 1986, 1987; Petuch and Sargent, 2012, for more detailed information on the molluscan assemblages of these environments). Some of the more important provinciatonal endemics that occur in these habitats include the following (several shown in Figures 3.5 and 3.6):

Gastropoda
Neritidae
   Nerita (Theliostyla) lindae
Cerithiidae
   Cerithium lindae
Strombidae
   Eustrombus gigas verrilli (a northern form or subspecies)
Modulidae
   Modulus pacei
Muricidae
   Stramonita buchecki (see Appendix 2)
Melongenidae
   Melongena (Rexmela) corona winnerae
Bivalvia
Veneridae
   Mercenaria hartae (see Appendix 2)

Figure 3.5 Gastropods of the Palm Beach Provinciatone coastal lagoons: (A, B) *Cerithium lindae* Petuch, 1987, length 12.3 mm. (C) *Modulus pacei* Petuch, 1987, width 14.5 mm. (D) *Jaspidiconus pfluegeri* Petuch, 2004, length 21.5 mm. (E, F) *Nerita (Theliostyla) lindae* Petuch, 1988, width 14.6 mm.

The largest of the coastal lagoon systems of the Palm Beach Provinciatone is Lake Worth in central Palm Beach County. Having mud and sand flats, mangrove forests, turtle grass beds, and oyster bioherms, Lake Worth houses all six of these lagoonal endemic gastropods. The widespread Georgian Subprovince cone shell *Jaspidiconus pfluegeri* is also common in shallow sand areas of Lake Worth. In true provinciatonal fashion, the Lake Worth lagoon is also one of the only places in the southern part of the Carolinian Province where Caribbean and Carolinian strombids occur sympatrically. Here, particularly around Peanut Island in northern Lake Worth, the Caribbean *Strombus pugilis* occurs along with the Carolinian *Strombus alatus*, with both species often forming immense intermixed aggregations. These small strombids occur together with Verrill's queen conch (*Eustrombus gigas verrilli*), an endemic northern form or subspecies of the Caribbean queen conch (*Eustrombus gigas*), which ranges from Fort Pierce south to Lake Worth (see Petuch and Sargent, 2012, for illustrations and discussions of Verrill's queen conch). Mollusks endemic to the Palm Beach Provinciatone deep offshore reefs are as follows (several shown in Figure 3.7):

Triviidae
 Pusula juyingae
Muricidae
 Dermomurex (Trialatella) glicksteini (Note: the *D. glicksteini* reported from Bermuda is a
  new, undescribed endemic species)
 Pygmaepterys richardbinghami

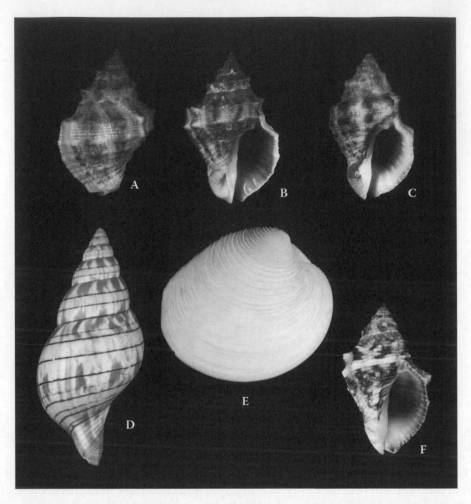

Figure 3.6 New discoveries from the Georgian Subprovince: (A, B) *Stramonita buchecki* Petuch, new species, holotype, length 28 mm. (C) *Stramonita rustica* (Lamarck, 1822), length 27 mm (specimen from Singer Island, Lake Worth Lagoon; for comparison with *S. buchecki*). (D) *Cinctura hunteria keatonorum* Petuch, new species, holotype, length 92 mm. (E) *Mercenaria hartae* Petuch, new species, holotype, width 44.28 mm. Confined to the coastal lagoons of the Palm Beach Provinciatone. (F) *Stramonita floridana* (Conrad, 1837), length 39.5 mm (specimen from Singer Island, Lake Worth Lagoon; for comparison with *stramonita buchecki*). The descriptions of these new taxa are given in Appendix 2.

Marginellidae
    Prunum evelynae (live specimen shown here on Figure 3.8)
    Prunum nobilianum
Cancellariidae
    Tritonoharpa janowskyi
Conidae
    Dauciconus glicksteini
    Gradiconus patglicksteinae
    Kellyconus binghamae
    Tuckericonus flamingo

Figure 3.7 Gastropods of the Palm Beach Provinciatone offshore deep reefs: (A) *Tuckericonus flamingo* (Petuch, 1980), holotype, length 19.3 mm. (B) *Kellyconus binghamae* (Petuch, 1987), holotype, length 17.4 mm. (C) *Duuciconus glicksteini* (Petuch, 1987), holotype, length 20.7 mm. (D) *Gradiconus patglicksteinae* (Petuch, 1987), holotype, length 25.5 mm. (E) *Pygmaepterys richardbinghami* (Petuch, 1987), holotype, length 16 mm. (F) *Tritonoharpa janowskyi* Petuch and Sargent, 2011, holotype, length 15.8 mm.

The best developed of these offshore coral reefs and adjacent sandy areas are found off Palm Beach Island, central Palm Beach County, where three separate, consecutively deeper reef tracts (coral bioherms) have developed in a step-like fashion. On the deepest of these reefs, in around 20–30 m depths, local divers have collected several rare Caribbean species on a regular basis (Petuch and Sargent, 2012). Some of these include the glory-of-the-Atlantic cone *Atlanticonus granulatus* (Conidae) and the Suriname cowrie *Propustularia surinamensis* (Cypraeidae).

The eastern Florida section of the Georgian Subprovince, from Palm Beach in the south to St. Augustine in the north, houses a distinctive set of populations of the crown conch, *Melongena (Rexmela) corona* (Melongenidae; Figure 3.9). This classic Florida species ranges from St. Augustine, around Florida Bay, and northward to Mobile Bay, Alabama, and has evolved six different varieties or subspecies that occur in separate areas of the Suwannean, Floridian, and southern Georgian Subprovinces (Petuch, 2004; Petuch and Sargent, 2012). Three of these subspecies (shown here in Figure 3.9) are restricted to intertidal lagoonal environments along eastern and northeastern Florida and include the following: *Melongena (Rexmela) corona sprucecreekensis* (St. Augustine to Daytona Beach), *Melongena (Rexmela) corona altispira* (Cape Canaveral to Stuart), and *Melongena (Rexmela)*

Figure 3.8 Close-up of a living specimen of the marginellid *Prunum evelynae* (Bayer, 1943), a species that is endemic to the Palm Beach Provinciatone of the Georgian Subprovince. (Courtesy of Marvin Glickstein, 1985.)

corona winnerae (St. Lucie River Mouth to southern Lake Worth; see Tucker, 1994, for a review of the biogeographical patterns and genetic differences between these subspecies and the western Florida subspecies). The evolution and ecological preferences of these three Georgian Subprovince subspecies is discussed and outlined in Petuch (2004), and these findings support Tucker's original insightful work.

Figure 3.9 Crown conch subspecies of the Georgian Subprovince: (A) *Melongena (Rexmela) corona sprucecreekensis* Tucker, 1994, length 140 mm. (B) *Melongena (Rexmela) corona altispira* Pilsbry and Vanatta, 1934, length 68 mm. (C) *Melongena (Rexmela) corona winnerae* Petuch, 2004, length 102 mm.

chapter four

Molluscan biodiversity in the subprovinces of the Florida Peninsula

The southern and western sections of the Florida Peninsula, including the Florida Keys, contain two separate faunal areas; the Floridian Subprovince (named for Florida) and the Suwannean Subprovince (named for the Suwannee River, which empties into the Gulf of Mexico in northwestern Florida) (Petuch, 2004). Both of these subprovinces are under the influence of very different physical oceanographic conditions, and this is reflected in the great differences in the compositions of their malacofaunas. The Floridian Subprovince is completely within the warm tropical water mass of the Florida Current, which flows northward through the Yucatan Straits and then eastward through the Straits of Florida and along the Florida Keys. The Suwannean Subprovince, on the other hand, is under the influence of the eastern side of the Gulf Loop Current, which brings cold water southward along western Florida during the winter months. The oceanographic conditions of the shallow Neritic Zone of the Suwannean Subprovince fluctuate seasonally, with cold (often below 20 degrees Celsius) temperatures in the winter and hot, tropical (above 25 degrees Celsius) in the summer. In direct contrast, the Floridian Subprovince, particularly the Florida Keys area, retains almost constant tropical–subtropical water conditions during the entire year, with temperatures never falling below 20 degrees Celsius.

Molluscan biodiversity in the Floridian Subprovince

Extending from approximately Fort Lauderdale, Broward County, Florida and along Biscayne Bay and the Florida Keys (Figure 4.1), to Naples, Collier County, Florida, the Floridian Subprovince is one of the two eutropical components of the Carolinian Province (the other is the Yucatanean Subprovince; see Chapter 5). Traditionally, this subprovince had been considered to be part of the old Caribbean Province of Briggs (1995), Warmke and Abbott (1961), and Woodward (1856) and the Gulf Province by Valentine (1973), but new, more detailed faunal analyses demonstrate that the area belongs within the Carolinian Province. The Floridian Subprovince contains all the classic Carolinian index taxa that were listed in Chapter 2 (and illustrated in Figures 2.2 and 2.3) but also exhibits a high degree of endemism that can be used to differentiate this area from the other Carolinian Subprovinces. As a discrete biogeographical entity, the boundaries of the Floridian Subprovince can be defined by the percentages of endemism of the 10 key index taxa. Based on the relationship

$$t = \frac{n}{N}(100), \, t > 25$$

these are as follows:

Modulidae ($t1$), $t = 100$, with $N = 1$, $n = 1$
Turbinellidae ($t2$), $t = 0$, with $N = 1$, $n = 0$
Conidae ($t3$), $t = 20$, with $N = 15$, $n = 3$

Figure 4.1 View of Missouri Key, Florida Keys, showing the typical coral limestone shoreline environment of the upper and central Florida Keys. In the background, note the large waterspout that has formed on the right side of the offshore thunderstorm. (From Petuch, E.J., *The Geology of the Florida Keys and Everglades*, Thomson Publishers, Mason, Ohio, 2008.)

Conilithidae (*t*4), *t* = 29, with *N* = 7, *n* = 2
Muricinae (*t*5), *t* = 33, with *N* = 18, *n* = 6
Fasciolariinae (*t*6), *t* = 33, with *N* = 3, *n* = 1
Lyriinae (*t*7), *t* = 100, with *N* = 1, *n* = 1
Olivinae (*t*8), *t* = 0, with *N* = 1, *n* = 0
Cancellariinae (*t*9), *t* = 25, with *N* = 4, *n* = 1
Plesiotritoninae (*t*10), *t* = 0, with *N* = 1, *n* = 0

The raw data for these analyses are taken from the species lists shown in Appendix 1. The subprovincial combined index, incorporating the endemicity of these 10 families and subfamilies and defined by the relationship

$$S = \sum_{n=1}^{10} \frac{tn}{10}, \ S > 25$$

yields the index quantity *S* = 39. This high level of subprovincial endemism (with *S* > 25) demonstrates that the Floridian Subprovince is a strong, differentiable biogeographical entity.

The Floridian Subprovince contains several very distinctive and characteristic habitat types, several of which are unique within the tropical western Atlantic. Primary among these are the extensive turtle grass (*Thalassia testudinum*), mangrove, and oyster bioherm environments of Florida Bay (the broad, shallow bay between the Florida mainland and the Florida Keys island chain); the sponge bioherms growing on exposed limestone sea floors in southern Florida Bay; the coral reef systems (both zonated reefs and coral bioherms) of the Florida Keys reef tract; and the deep reefs and deep terraces along the Straits of Florida. These four habitat types, together, contain the bulk of the endemic species found in the Floridian Subprovince.

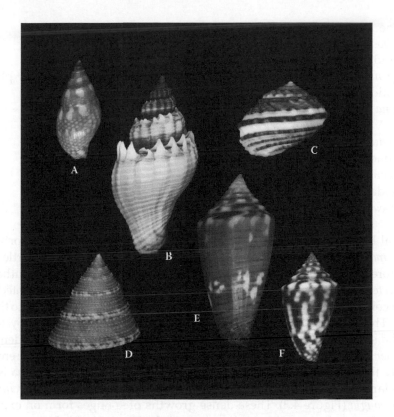

Figure 4.2 Gastropods endemic to the Florida Bay sea grass environments: (A) *Zaphrona taylorae* Petuch, 1987, length 10 mm. (B) *Melongena (Rexmela) bicolor* (Say, 1827), length 32.4 mm. (C) *Modulus calusa* Petuch, 1988, width 12 mm. (D) *Calliostoma adelae* Schwengel, 1951, height 15.8 mm. (E) *Gradiconus burryae* (Clench, 1942), length 35.5 mm. (F) *Jaspidiconus pealii* (Green, 1830), length 13.2 mm.

Florida Bay ecosystems

Being a wide, shallow embayment with permanent eutropical conditions, this geographical subdivision of southernmost Florida is unique within the Carolinian Province. The most notable and prominent ecosystem within the bay is the one centered on turtle grass beds, which cover at least 75% of the exposed sea floor. These immense sea grass beds are best developed behind the upper and middle Keys, such as Key Largo, Upper Matecumbe Key, and Vaca Key, and also in the very shallow central areas of the bay, such as near Rabbit Key Basin (see Petuch and Sargent, 2012, for more detailed information). Although they contain a typical tropical western Atlantic *Thalassia*-associated molluscan fauna, several characteristic endemic gastropods are also present in the turtle grass biomes and include the following (several illustrated in Figure 4.2):

Calliostomatidae
  Calliostoma adelae
Turbinidae
  Astraea (Lithopoma) americana
Modulidae
  Modulus calusa

Buccinidae
  Gemophos pacei (on oyster biostromes along northern Florida Bay)
Melongenidae
  Melongena (Rexmela) bicolor (also found in tide pools on exposed rocky platforms)
Columbellidae
  Zaphrona taylorae
Conidae
  Gradiconus burryae
Conilithidae
  Jaspidiconus pealii
Haminoeidae
  Haminoea taylorae

In central Florida Bay and Black Water Sound (behind Key Largo), Taylor's bay scallop *Argopecten irradians taylorae* typically forms immense shoals within the turtle grass beds. Although more typical of the Suwannean Subprovince coastal areas, this southern member of the *Argopecten irradians* complex ranges into Florida Bay, where it is represented by a dwarf, often highly colored, variety. Along the northern edge of Florida Bay, adjacent to the mainland and the Ten Thousand Islands mangrove keys, large *Crassostrea rhizophorae* oyster bioherms dominate the shallow intertidal areas. These biostromes also house an endemic buccinid gastropod, *Gemophos pacei*, which is unique to the mangrove jungle and oyster environments.

Probably the most distinctive marine ecosystem in the Florida Bay area is one that is centered on large bioherms composed primarily of sponges in the genera *Ircinia*, *Haliclona*, and *Spheciospongia* (Figure 4.3). These dense growths of sponges form on exposed oölitic limestone sea floors all along the northern side of the lower Florida Keys, where they live attached directly to the rocky substrate. The entire limestone sea floor is covered by a thin layer of coarse carbonate sand, which houses a cryptic but highly endemic malaco-fauna. The sponges and their associated mollusks are intermingled with dense growths of dasycladacean calcareous algae such as *Halimeda*, *Penicillus*, and *Acetabularia* and small bioherms of the scleractinian corals *Porites* and *Siderastrea*, making this one of the richest and most interesting marine communities in the Floridian Subprovince. Of special interest here is the small, slender endemic conid *Gradiconus mazzolii* (Figures 4.3 and 4.4D), which closely resembles *Gradiconus scalaris* from the Eastern Pacific Panamic Province (Petuch and Sargent, 2011a) and the endemic dwarf bubble shell, *Bulla striata frankovichi* (Figure 4.4B), which is characteristically marked with a color pattern of four dark brown bands (Petuch and Sargent, 2011a). During just one summer of fieldwork in this sponge-based ecosystem (2011), three new gastropods were discovered, which demonstrates how unexplored many of the shallow water areas in Florida Bay really are. Some of the endemic gastropods found in the sponge bioherms include the following (with several shown in Figure 4.4):

Nassariidae
  Nassarius websteri
Muricidae
  Vokesimurex rubidus marcoensis (may be a full, valid endemic species)
Busyconidae
  Fulguropsis spiratum keysensis (see Appendix 2)
Conidae
  Gradiconus mazzolii

Figure 4.3 View of the oölitic limestone sea floor, 2 m depth, off Little Torch Key, lower Florida Keys, showing the sponge *Haliclona*, the sea star *Echinaster*, and the endemic cone shell *Gradiconus mazzolii*. Mazzoli's cone is occupied by the shell-dwelling sipunculid worm *Phascolion cryptum*, which has extended its evert proboscis. (Photograph courtesy of N. Mazzoli, 2011.)

Drilliidae
   Cerodrillia clappi
   Neodrillia blacki
Raphitomidae
   Pyrgocythara hemphilli
Bullidae
   Bulla striata frankovichi (may be a full, distinct endemic species)

On the exposed limestone platforms along the shorelines of the Florida Keys, the characteristic endemic melongenid *Melongena (Rexmela) bicolor* (Figure 4.2B) is abundant in tide pools on the ocean sides of the islands. Here, this Keys endemic predatory gastropod feeds on dense aggregations of the batillariid *Batillaria minima* and the cerithiid *Cerithium lutosum*.

Florida Keys reef tracts

Extending from Key Biscayne, Miami, all the way to the Dry Tortugas, the Floridian Subprovince reef tract is the largest living coralline structure found in the United States. This extensive reef system follows the template that was established by the deposition of coral reef tracts during the Pliocene and Pleistocene and is essentially an extension of the carbonate deposition that began in the Okeechobean Sea of the Caloosahatchian Province (see Petuch, 2004; Petuch and Roberts, 2007, for details on the formation of ancient Floridian reef systems). Today, the main reef systems of the Florida Keys reef tract are

Figure 4.4 Gastropods of the oolitic limestone sea floors of Florida Bay: (A) *Neodrillia blacki* Petuch, 2004, length 23.4 mm. (B) *Bulla striata frankovichi* Petuch and Sargent, 2011, holotype, length 10.2 mm. (C) *Nassarius websteri* Petuch and Sargent, 2011, holotype, length 8 mm. (D) *Gradiconus mazzolii* Petuch and Sargent, 2011, length 20.8 mm. (E) *Fulguropsis spiratum keysensis* Petuch, new subspecies, holotype, length 57.8 mm (see Appendix 2). (F) *Vokesimurex rubidus marcoensis* (Sowerby, 1900), length 28.4 mm.

roughly 5 miles (7 km) offshore of the islands of the Florida Keys and are separated from the island mainlands by the wide, shallow Hawk Channel. The open sandy bottoms, scattered turtle grass beds, and patch reef bioherms of the Hawk Channel and the living coral reefs of the main tract together house the richest molluscan fauna known in the contiguous United States. Reef-associated endemic species also make up a significant part of the Keys malacofauna, with several belonging to genera that are found nowhere else in the Atlantic Ocean. Some of these, such as *Cyclothyca* (Amalthinidae) and *Quoyula* (Muricidae-Magilinae), are widespread components of the Eastern Pacific Panamic faunas, and their only known occurrence in the western Atlantic is in the Florida Keys region (Petuch, 1987; Petuch and Sargent, 2012). A rich and distinctive fauna of conilithids is also present on the reef tract, with the bright pink *Jaspidiconus fluviamaris* (Figure 4.5C) being one of the most distinctive and interesting species (Petuch and Sargent, 2011b). The following is a list of some of the more prominent and important endemic gastropods of the Florida Keys reef tract and adjacent areas (some shown here in Figures 4.5, 4.6, and 4.7):

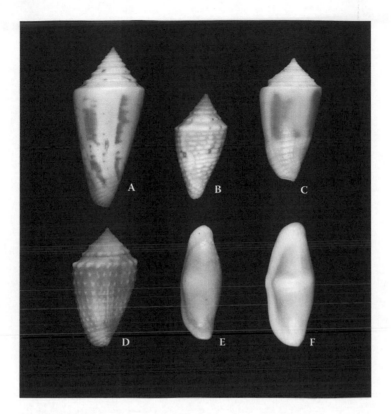

Figure 4.5 Endemic gastropods of the Florida Keys reef tract: (A) *Gradiconus anabathrum tranthami* (Petuch, 1995), length 26 mm (*G. antoni* Cargile and *tortuganus* Petuch and Sargent are synonyms; typical slender variety from Pickles Reef). (B) *Jaspidiconus acutimarginatus* (Sowerby, 1866), length 13 mm. (C) *Jaspidiconus fluviamaris* Petuch and Sargent, 2011, holotype, length 16.2 mm. (D) *Jaspidiconus vanhyningi* (Rehder, 1944), length 16 mm. (E) *Cyphoma sedlaki* Cate, 1976, length 13.5 mm. (F) *Cyphoma rhomba* Cate, 1978, length 16.9 mm.

Turbinidae-Liotiinae
  Arene vanhyningi
Ovulidae
  Cyphoma alleneae (possibly also in Cuba and Cay Sal Bank)
  Cyphoma rhomba
  Cyphoma sedlacki
Muricidae
  Dermomurex pacei (possibly also in northern Cuba)
  Favartia pacei
  Murexiella kalafuti
  Quoyula kalafuti
Volutidae
  Enaeta cf. cylleniformis (Biscayne Bay and northern Florida Keys; possibly a new species)
  Scaphella junonia elizabethae (shallow water subspecies; 3–10 m depths)

Figure 4.6 Endemic gastropods of the Florida Keys reef tract: (A) *Scaphella junonia elizabethae* Petuch and Sargent, 2011, length 65 mm. (B) *Murexiella kalafuti* Petuch, 1987, holotype, length 15.4 mm. (C) *Cyclothyca pacei* Petuch, 1987, length 9.5 mm. (D) *Quoyula kalafuti* Petuch, 1987, holotype, length 12.9 mm. (E) *Favartia pacei* Petuch, 1988, length 15.6 mm. (F) *Dibaphimitra florida* (Gould, 1856), length 46.5 mm.

Conidae
  Gradiconus anabathrum tranthami (= *antoni* Cargile, 2011)
Conilithidae
  Jaspidiconus acutimarginatus
  Jaspidiconus fluviamaris (also in the southern part of the Palm Beach Provinciatone)
Drilliidae
  Drillia moseri
Amalthinidae
  Cyclothyca pacei

Deep-water areas off the Florida Keys

Along the ocean side of the Florida Keys reef tract, the continental shelf is extremely narrow, plummeting directly into the deep water of the Straits of Florida. A series of terraces, cut by Pleistocene low sea levels, edge this continental margin and form a unique set of habitats just offshore of the reef systems. These deep-water terraces (100–300 m depths) are virtually unexplored, but the little research that has been conducted in these areas has shown that a highly endemic malacofauna exists all along the Florida Keys. Of special interest in these deep-water areas is a large endemic species radiation of the

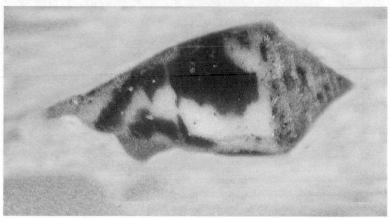

Figure 4.7 Close-up of living specimens of endemic Florida Keys *Gradiconus* species. Top: Mazzoli's cone shell, *Gradiconus mazzolii* Petuch and Sargent, 2011, from the sponge bioherms off Middle Torch Key, lower Florida Keys. Note the yellow color of the living animal. This newly discovered species is endemic to shallow-water limestone sea floors along the Florida Bay side of the lower Florida Keys. Bottom: Trantham's cone shell, *Gradiconus anabathrum tranthami* (Petuch, 1995), from a coral reef off the Atlantic Ocean side of Missouri Key, middle Florida Keys. Note the pale pink-orange color of the living animal. This variably colored Keys subspecies of the western Florida *Gradiconus anabathrum* (Crosse, 1865) lives in clean coral sand pockets on living coral reefs, from Key Biscayne south to the Dry Tortugas. The closely related Keys endemic *Gradiconus burryae* (Clench, 1942), which lives in turtle grass beds, has a dark red or reddish-purple animal. (Photographs courtesy of William Bennight.)

genus *Inodrillia* (Drilliidae), comprising at least six species (see Tucker, 2004). Some of the Floridian Subprovince deep-water endemic gastropods include the following (several illustrated here in Figure 4.8):

Calliostomatidae
   Calliostoma frumari
Ovulidae
   Pseudocyphoma gibbulum (also found in the Palm Beach Provinciatone)

Figure 4.8 Deep-water endemic gastropods of the Florida Keys area: (A) *Prunum frumari* Petuch and Sargent, 2011, holotype, length 12 mm. (B) *Pseudocyphoma gibbulum* (Cate, 1978), length 13.5 mm. (C) *Scaphella (Clenchina) dohrni* (Sowerby III, 1903) form *florida* (Clench and Aguayo, 1940), length 57 mm. (D) *Aurinia (Rehderia) schmitti* (Bartsch, 1931), length 132 mm. (E) *Bartschia frumari* Garcia, 2008, length 31 mm. (F) *Prunum redfieldi* (Tryon, 1882), length 12 mm.

Muricidae
    Pterynotus guesti
Buccinidae
    Bartschia frumari
    Ptychosalpinx globulus
Volutidae
    Aurinia (Rehderia) schmitti (also found north of the Dry Tortugas)
    Scaphella (Clenchina) dohrni (and stepped-spired form *florida*)
Olivellidae
    Olivella moorei
Marginellidae
    Prunum frumari
    Prunum redfieldi
Terebridae
    Strioterebrum rushii

Drilliidae
  Inodrillia dido
  Inodrillia hesperia
  Inodrillia hilda
  Inodrillia ino
  Inodrillia miamia
  Inodrillia vetula

One of the most important collectors of deep-water Florida Keys shells is Frank Frumar of Cudjoe Key, who brought to light many rarely seen species while conducting deep-water dredging off the Lower Keys during the 1990s and early 2000s (from 100 to 350 m depths). In addition to rediscovering many rare deep-water Keys endemics, such as the marginellid *Prunum redfieldi* (Figure 4.8F), Frumar also dredged up several important new species from these unexplored deep-water environments. Some of these Keys endemics include *Calliostoma frumari*, *Bartschia frumari*, and *Prunum frumari* (see Figure 4.8).

Molluscan biodiversity in the Suwannean Subprovince

Extending from Naples, Collier County, Florida, in the south to Mobile Bay, Alabama, in the north, the Suwannean Subprovince represents a paratropical subdivision of the Carolinian Province. This eastern Gulf of Mexico faunal component was included in Briggs's Gulf Province (1995), but the presence of a full compliment of classic Carolinian index taxa demonstrates that this area belongs within the boundaries of the Carolinian Province. As a discrete biogeographical entity, the Suwannean Subprovince can be defined by the percentages of endemism in the 10 key index taxa. Based on the relationship

$$t = \frac{n}{N}(100), t > 25$$

these are as follows:

  Modulidae ($t1$), $t = 100$, with $N = 1$, $n = 1$
  Turbinellidae ($t2$), $t = 0$, with $N = 1$, $n = 0$
  Conidae ($t3$), $t = 33$, with $N = 9$, $n = 3$
  Conilithidae ($t4$), $t = 25$, with $N = 4$, $n = 1$
  Muricinae ($t5$), $t = 20$, with $N = 15$, $n = 3$
  Fasciolariinae ($t6$), $t = 50$, with $N = 4$, $n = 2$
  Lyriinae ($t7$), $t = 0$, with $N - 0$, $n = 0$ (absent in the Suwannean Subprovince)
  Olivinae ($t8$), $t = 66$, with $N = 3$, $n = 2$
  Cancellariinae ($t9$), $t = 25$, with $N = 4$, $n = 1$
  Plesiotritoninae ($t10$), $t = 0$, with $N = 1$, $n = 0$

The raw data for these analyses were taken from the species lists shown in Appendix 1. The subprovincial combined index, incorporating the endemicity of these 10 families and subfamilies and defined by the relationship

$$S = \sum_{n=1}^{10} \frac{tn}{10}, \ S > 25$$

Figure 4.9 Coastal Suwannean Subprovince endemic gastropods: (A) *Gradiconus anabathrum* (Crosse, 1865), length 43 mm. (B) *Gradiconus anabathrum* (Crosse, 1865) form *floridensis* (Sowerby, 1870), length 43.3 mm. (C) *Oliva (Americoliva) sayana sarasotaensis* Petuch and Sargent, 1986, length 48 mm. (D) *Jaspidiconus stearnsi* (Conrad, 1869), length 16.7 mm. (E) *Melongena (Rexmela) corona* (Gmelin, 1791), length 58 mm.

yields the index quantity $S = 31.9$. This high level of subprovincial endemism (with $S > 25$) demonstrates that the Suwannean Subprovince is a differentiable biogeographical entity.

Among the many habitats along the eastern Gulf of Mexico, the Suwannean Subprovince contains two environments that are of special interest: the quartz sand coastal lagoons, which are dominated by the emergent salt-tolerant grass *Spartina* and by bioherms and banks of the oyster *Crassostrea virginica*, and the deep-water (100–200 m depths) offshore calcareous algal beds, which are dominated by the coralline algae *Porolithon*. Both of these special habitats support rich malacofaunas with high levels of endemicity, including many relictual taxa. The shallow water lagoonal areas contain several important relicts of the mid-Pleistocene Bermont Formation (see Petuch, 2004; Petuch and Roberts, 2007), including the muricids *Vokesinotus perrugatus* and *Eupleura tampaensis* (both incorrectly placed in the genus *Urosalpinx* by Abbott, 1974). The classic Pliocene Caloosahatchian Province genus *Vokesinotus* was originally thought to have become extinct by the mid-Pleistocene but is now recognized as being extant within the Suwannean Subprovince. Some of the typical shallow lagoonal Suwannean Subprovince endemic species include the following (several illustrated in Figures 4.9 and 4.10):

Gastropoda
Calliostomatidae
   Calliostoma tampaense

Figure 4.10 Coastal Suwannean Subprovince endemic mollusks: (A) *Naticarius verae* Rehder, 1947, height 15 mm. (B) *Eupleura tampaensis* (Conrad, 1846), length 23 mm. (C) *Vokesinotus perrugatus* (Conrad, 1846), length 22 mm. (D) *Argopecten irradians taylorae* Petuch, 1987, length 34 mm. (also found in Florida Bay). (E) *Modulus floridanus* (Conrad, 1869), height 11 mm. (F) *Murexiella taylorae* Petuch, 1987, holotype, length 15.6 mm.

Potamididae
   Cerithideopsis costatus turritus
Modulidae
   Modulus floridanus
Naticidae
   Naticarius verae
Muricidae
   Eupleura tampaense
   Murexiella taylorae (offshore on deeper water oyster bioherms)
   Stramonita canaliculata (northwestern Florida to northeastern Texas)
   Vokesinotus perrugatus
Melongenidae
   Melongena (Rexmela) corona corona (Naples to Cedar Key)
   Melongena (Rexmela) corona johnstonei (Cedar Key to Mobile Bay, Alabama; see Tucker, 1994, for an overview of the western Florida corona subspecies)
Olividae
   Oliva (Americoliva) sayana sarasotaensis (offshore in deeper water)
Conidae
   Gradiconus anabathrum anabathrum

Figure 4.11 Endemic gastropods of the northeastern Gulf of Mexico: (A) *Cyphoma robustior* Bayer, 1941, holotype, length 37.6 mm. (B) *Oliva (Americoliva) sunderlandi* Petuch, 1987, holotype, length 22 mm. (C) *Melongena (Rexmela) corona johnstonei* Clench and Turner, 1956, length 92 mm. (D, E) *Scaphella junonia johnstoneae* Clench, 1953, length 119 mm. (F) *Hesperisternia grandana* (Abbott, 1986), length 27 mm.

Conilithidae
    Jaspidiconus stearnsi
Bivalvia
Pectinidae
    Argopecten irradians taylorae (also found in Florida Bay)

In the extreme northeastern section of the Gulf of Mexico, in the area between the Apalachicola River Delta and the mouth of the Suwannee River, the molluscan fauna has been shown to be of special significance. In the winter, these deeper water areas (20–100 m depths) remain at a much warmer temperature than do the shallow coastal waters, and this more stable environment has allowed for the evolution of a distinctive endemic para-tropical malacofauna. Some of these northeastern Gulf endemics include the following (several illustrated in Figure 4.11):

Cerithiidae
    Cerithium chara
    Cerithium lymani

Ovulidae
   Cyphoma robustior (may range to north of the Dry Tortugas; on antipatharian gorgonians)
Buccinidae
   Hesperisternia grandana (referred to as *Cantharus multangulus grandanus* by Abbott, 1986)
Volutidae
   Scaphella junonia johnstoneae (also ranges to off Alabama)
Olividae
   Oliva (Americoliva) Sunderlandi

Of the two areas of special ecological interest within the Suwannean Subprovince, the deep-water coralline algae beds contain the largest and most impressive endemic molluscan fauna. Extending from south of Apalachicola to north of the Dry Tortugas, these beds of *Porolithon* red calcareous algae hug the outer edge of the continental shelf, at depths of 35–200 meters, and form dense intertwined accumulations that are often several meters thick. Interestingly enough, the deeper part of this unusual Suwannean offshore environment is the only locality in the entire Carolinian Province where a shallow-water-derived species, *Lindafulgur lyonsi* (Busyconidae), lives sympatrically with a deep-water–derived species, *Perotrochus amabilis* (Pleurotomariidae). To find a shallow-water slit shell living together with a deep-water *Busycon* whelk is an amazing discovery and represents one of the most unusual bathymetric overlaps in the entire western Atlantic (William Lyons, personal communication; from data collected by the State of Florida *Hour Glass* fisheries research data). For more detailed data, illustrations, and species lists of the faunal components of this unique offshore ecosystem, see Petuch and Sargent (2012, Chapter 4).

Some of the endemic Suwannean mollusks found in these deep-water coralline algal beds include the following (several shown in Figures 4.12 and 4.13):

Gastropoda
Pleurotomariidae
   Perotrochus amabilis (associated with sponges growing on rhodolith beds)
Turritellidae
   Torculoidella lindae
Ovulidae
   Cyphoma lindae (associated with antipatharian gorgonians on rhodolith beds)
Muricidae
   Chicoreus rachelcarsonae
   Favartia lindae
   Phyllonotus pomum whymani
   Vokesimurex lindajoyceae
Fasciolariidae
   Cinctura tortugana
   Fasciolaria bullisi (also found along the northern Gulf of Mexico and Campeche Banks)
   Fusinus stegeri
   Heilprinia lindae
   Leucozonia jacarusoi
Busyconidae
   Lindafulgur lyonsi

Figure 4.12 Endemic mollusks of the deep-water calcareous algal beds: (A) *Vokesimurex lindajoyceae* (Petuch, 1987), length 32.4 mm. (B) *Myurellina lindae* Petuch, 1987, holotype, length 64 mm. (C) *Dauciconus aureonimbosus* (Petuch, 1987), holotype, length 25.5 mm. (D) *Lindafulgur lyonsi* (Petuch, 1987), holotype, length 128 mm. (E) *Cinctura tortugana* (Hollister, 1957), length 70.8 mm. (F) *Lindapecten lindae* Petuch, 1995, length 16 mm.

Buccinidae
    Hesperisternia harasewychi
Volutidae
    Scaphella (Caricellopsis) matchetti
    Scaphella (Clenchina) marionae
Cancellariidae
    Cancellaria richardpetiti
Conidae
    Dauciconus aureonimbosus
    Lindaconus aureofasciatus
Terebridae
    Myurellina lindae
Turridae
    Polystira sunderlandi
Drilliidae
    Fenimorea sunderlandi (may also occur off Honduras)
Bivalvia

Figure 4.13 Endemic gastropods of the deep-water calcareous algal beds: (A) *Cancellaria richardpetiti* Petuch, 1987, length 47 mm. (B) *Scaphella (Curicellopsis) matchetti* Petuch and Sargent, 2011, holotype, length 35.9 mm. (C) *Favartia lindae* Petuch, 1987, length 18 mm. (D) *Hesperisternia harasewychi* (Petuch, 1987), holotype, length 24.8 mm. (E) *Chicoreus rachelcarsonae* Petuch, 1987, holotype, length 40 mm. (F) *Cyphoma lindae* Petuch, 1987, holotype, length 17 mm.

Pectinidae

Lindapecten lindae (Note: Mikkelsen and Bieler [2007] incorrectly placed the closely related *Lindapecten muscosus* in the genus *Aequipecten* and ignored the genus *Lindapecten*. This is now known to be in error, as demonstrated by Waller [2011], who fully accepts the genus *Lindapecten* as the valid name for this group of endemic western Atlantic spiny scallops. The type species of *Lindapecten* is *L. lindae*.)

Further research and collecting may show that this deep-water area contains the richest and most ecologically diverse invertebrate fauna in the entire Gulf of Mexico. It is here hoped that petroleum derivatives from the 2010 *Deep Horizon* oil spill have not harmed this U.S. biological treasure.

chapter five

Southern and western subprovinces of the Carolinian Province

Unlike the Caribbean and Brazilian Provinces, which are contiguous marine biogeographical entities, the Carolinian Province is divided into two separate sets of subprovinces. The bisected faunal pattern in the Gulf of Mexico is caused by the ecological barrier that is formed by the immense freshwater effluent of the wide Mississippi River Delta. This physical and ecological separation from the Suwannean Subprovince has resulted in the genetic isolation of the molluscan faunas of the western Texan and southern Yucatanean Subprovinces. Although they contain the full complement of Carolinian index taxa listed in Chapter 2 (and illustrated in Figures 2.2 and 2.3), the Texan and Yucatanean Subprovinces also have evolved their own distinct endemic faunas.

Molluscan biodiversity in the Texan Subprovince

Extending from the Mississippi River Delta westward to Veracruz, Veracruz State, Mexico, the Texan Subprovince (named for the State of Texas; Petuch, 2004) contains an impoverished molluscan fauna. Several ecological and climatic factors contribute to this lack of species richness and diversity, primarily because of the cold winter water temperatures produced by the westward-flowing northern section of the Gulf Loop Current. These winter temperate water conditions (often falling below 20 degrees Celsius) are reversed in the summer months, when water temperatures reach high tropical levels (often over 30 degrees Celsius). The ecological stress on marine organisms produced by these wildly fluctuating water temperatures is further compounded by fluctuating salinities in the coastal lagoons and bays. During dry periods in the late summer and autumn, these areas develop hypersaline conditions, with salinities often exceeding 50 parts per thousand. During the high precipitation periods of the spring and early summer, the coastal lagoons and bays become inundated with fresh water and develop hyposaline conditions, with salinities of often less than 20 parts per thousand. These widespread highly variable ecological conditions restrict the local malacofaunas to a small number of physiologically plastic species. For a detailed overview of the Texan marine habitats and local ecosystems, see Tunnel et al. (2010).

The Texan Subprovince, as defined here, was originally considered to be the western part of Briggs's (1995) Gulf Province. As a discrete biogeographical entity, the boundaries of the Texan Subprovince can be defined by the percentages of endemism in the 10 key index taxa.

Based on the relationship

$$t = \frac{n}{N}(100), \; t > 25$$

these are as follows:

Modulidae (*t*1), *t* = 100, with *N* = 1, *n* = 1
Turbinellidae (*t*2), *t* = 0, with *N* = 0, *n* = 0 (absent from the Texan subprovince)
Conidae (*t*3), *t* = 11, with *N* = 9, *n* = 1
Conilithidae (*t*4), *t* = 40, with *N* = 5, *n* = 2
Muricinae (*t*5), *t* = 0, with *N* = 10, *n* = 0
Fasciolariinae (*t*6), *t* = 33, with *N* = 3, *n* = 1
Lyriinae (*t*7), *t* = 0, with *N* = 0, *n* = 0 (absent from the Texas subprovince)
Olivinae (*t*8), *t* = 50, with *N* = 2, *n* = 1
Cancellariinae (*t*9), *t* = 50, with *N* = 2, *n* = 1
Plesiotritoninae (*t*10), *t* = 0, with *N* = 1, *n* = 0

The raw data for these analyses were taken from the species lists in Appendix 1. The subprovincial combined index, incorporating the endemicity of these 10 families and subfamilies and defined by the relationship

$$\sum_{n=1}^{10} \frac{tn}{10}, \ S > 25$$

yields the index quantity *S* = 28.4. This high level of subprovincial endemism (*S* > 25) demonstrates that the Texan Subprovince is a differentiable biogeographical entity.

Of special interest within the Texan Subprovince is the coral bioherm-associated molluscan fauna found on the East and West Flower Garden Banks, approximately 110 miles (175 km) southeast of Galveston, Texas (Bright and Pequegnat, 1974; Rezak, Bright, and McGrail, 1985). These carbonate environment offshore banks, bathed by the warm northward-flowing post-Yucatan component of the Gulf Loop Current, are considered to be the farthest north coral biohermal structures in the Gulf of Mexico. Over 200 species of eutropical mollusks have been collected here, including such classic Caribbean Province index species as *Taenioturbo cailleti* and *Taenioturbo canaliculatum* (both Turbinidae), *Eustrombus gigas* and *Aliger gallus* (both Strombidae), *Babelomurex scalariformis* (Muricidae: Magilinae), and *Stephanoconus regius* and *Gladioconus mus* (both Conidae). Since no endemic species are known from this area, the Flower Garden Banks do not appear to be a provinciatone between the Caribbean and Carolinian Provinces.

The far easternmost edge of the Texan Subprovince, encompassing the Mississippi River Delta, the western Louisiana coast, and the northeastern coast of Texas as far as Port Arthur, constitutes a separate section of the subprovincial area. Here, the deeper water sea floors (100–300 m depths) are dominated by muddy substrates that support only an impoverished molluscan fauna. Of primary interest on these deltaic offshore environments is a small but important species radiation of the family Volutidae, including *Aurinia kieneri ethelae* (Figure 5.1A,B), an endemic Mississippi Delta subspecies of the widespread Gulf of Mexico *Aurinia kieneri* characterized by having large undulating longitudinal folds on the body whorl, and *Scaphella (Clenchina) worki*, an unusual species with highly stepped scalariform spire whorls. Occurring with these distinctive endemic volutes are several western Gulf of Mexico endemic cone shells, including *Conasprelloides cancellatus* (= *C. austini*; Conidae) and *Dalliconus armiger* (= *D. clarki* and *D. frisbeyae*; Figure 5.1E) and *D. sauros* (both Conilithidae; Figure 5.1C,D), and the Delta endemic *Cancellaria rosewateri* (Cancellariidae). These deeper water Mississippi Delta endemic species may be endangered or now extinct due to the massive amounts of petroleum-related toxins produced by the British Petroleum 2010 *Deep Horizon* oil spill.

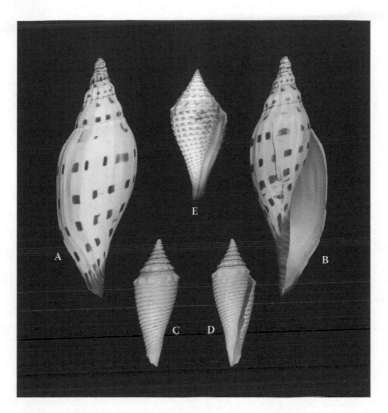

Figure 5.1 Mississippi River Delta deep-water gastropods: (A, B) *Aurinia kieneri ethelae* Pilsbry and Olsson, 1953, length 111 mm (note the large longitudinal folds that are characteristic of this subspecies). (C, D) *Dalliconus sauros* (Garcia, 2006), holotype, length 29.5 mm. (E) *Dalliconus armiger* (Crosse, 1858), length 32.5 mm (type of the synonym *frisbeyae*).

Texan coastal lagoons

The Texas coastline, from near Galveston Bay south to Laguna Madre on the Mexico border, is physiographically dominated by barrier island systems and extensive coastal lagoons. As described earlier, the salinity and water temperatures in these lagoons fluctuate seasonally, and these harsh ecological parameters result in an impoverished, low-diversity malacofauna. Although they have a low species richness, these lagoonal molluscan assemblages contain several distinctive endemic species and subspecies, including the following (illustrated in Figure 5.2):

Gastropoda
Ovulidae
  Simnialena marferula
Busyconidae
  Fulguropsis plagosus galvestonensis
  Fulguropsis plagosus texanum
  Sinistrofulgur pulleyi
Olividae
  Oliva (Americoliva) sayana texana

Figure 5.2 Texan Subprovince coastal lagoon mollusks: (A) *Argopecten irradians amplicostatus* (Dall, 1898), length 36 mm. (B) *Oliva (Americoliva) sayana texana* Petuch and Sargent, 1986, length 52 mm. (C) *Oliva (Americoliva) sayana texana* Petuch and Sargent, 1986, length 58 mm. (D) *Gradiconus largillierti* (Kiener, 1845), length 39 mm.

Conidae
  Gradiconus largillierti (possibly also on the Campeche Banks)
Terebridae
  Strioterebrum texanum
Bivalvia
Pectinidae
  Argopecten irradians amplicostatus (also the northern Yucatanean Subprovince)
Veneridae
  Mercenaria texana

The classic Carolinian Province olive shell *Oliva (Americoliva) sayana* is actually represented by three geographically separated allopatric subspecies: *O. sayana texana* (Figure 5.2B and C), which ranges from Tampico, Mexico, to the Mississippi River Delta; *O. sayana sarasotaensis* (Figure 4.9C), which ranges from the Mississippi coast to Naples, Florida; and *O. sayana sayana* (Figure 2.2), which ranges from Palm Beach, Florida, to Cape Hatteras, North Carolina (Petuch and Sargent, 1986). The Texas subspecies is physically separated from the western Florida subspecies by the wide, fresh and brackish water barrier of the Mississippi River Delta, while the eastern U.S. subspecies is separated from the Gulf of Mexico populations by the carbonate and coral reef environments of the Florida

Keys. This tripartite suite of allopatric subspecies is seen in other groups of mollusks, such as the *Argopecten irradians* species complex.

Molluscan biodiversity in the Yucatanean Subprovince

Extending from Tampico, Mexico, to Isla Contoy on the extreme northeastern tip of the Yucatan Peninsula of Mexico, the Yucatanean Subprovince (named for the Yucatan Peninsula; Petuch, 2004) contains one of the richest malacofaunas found in the entire Carolinian Province. Due to its proximity to the warm, northward-flowing lower section of the Gulf Loop Current, the Yucatanean Subprovince exhibits year-round eutropical oceanographic conditions and contains a highly endemic malacofauna. Like the offshore areas of the Suwannean Subprovince, the Yucatanean area also houses a large number of relictual genera and species complexes that are derived from Caloosahatchian Paleoprovince ancestors. Some of the more prominent of these include busyconid whelks of the genera *Lindafulgur* and *Busycuurctum* and the melongenid *Melongena (Rexmela) bispinosa*.

The Yucatanean Subprovince contains two environments that are of special interest and that contain the bulk of the endemic taxa: the tropical coastal lagoon systems (Figure 5.3), and the offshore Campeche Banks and the banks off Cabo Catoche. Unlike the coastal lagoons of Texas, which are dominated by organic-rich siliciclastic muds, the lagoons of the Yucatan Peninsula are dominated by carbonate muds (calcilutites) and turtle grass beds and sedimentologically closely resemble the carbonate mud areas of the Florida Keys. Likewise, the Campeche Banks, the nearby Alacran Reefs, and the deeper water areas off Cabo Catoche and Isla Contoy all are dominated by coarse carbonate sediments and living coral reefs and closely resemble the environments around the Florida Keys reef

Figure 5.3 View of a coastal lagoon in southern Tamaulipas State, Mexico, north of Tampico. This type of lagoon system is typical of the Mexican coastline along the extreme southern section of the Texan Subprovince and across the entire Yucatanean Subprovince. (From E. Petuch, archival photograph, 1973.)

tract. By having eutropical water conditions and widespread carbonate environments, the Yucatanean and Floridian Subprovinces can be considered sister biogeographical units, both separated by Cuba and the Antillean Subprovince of the Caribbean Province.

Because of the preponderence of resident eutropical Caribbean Province taxa, such as *Eustrombus gigas* (Strombidae), *Melongena melongena* (Melongenidae), *Turbinella angulata* (Turbinellidae), and *Stephanoconus regius* and *Gladioconus mus* (both Conidae), most marine molluscan biogeographers (Woodward, 1856; Warmke and Abbott, 1961; Valentine, 1973; Briggs, 1995) have considered the Yucatanean area to be part of the Caribbean Province. The data supplied in Appendix 1 show that the Yucatanean Subprovince is a discrete biogeographical entity within the Carolinian Province and that its boundaries can be defined by the percentages of endemism in the 10 key index taxa. Based on the relationship

$$t = \frac{n}{N}(100),\ t > 25$$

these are as follows:

Modulidae (*t1*), *t* = 100, with *N* = 1, *n* = 1
Turbinellidae (*t2*), *t* = 50, with *N* = 4, *n* = 2
Conidae (*t3*), *t* = 21, with *N* = 14, *n* = 3
Conilithidae (*t4*), *t* = 33, with *N* = 6, *n* = 2
Muricinae (*t5*), *t* = 19, with *N* = 16, *n* = 3
Fasciolariinae (*t6*), *t* = 25, with *N* = 4, *n* = 1
Lyriinae (*t7*), *t* = 0, with *N* = 0, *n* = 0 (absent from the Yucatanean Subprovince)
Olivinae (*t8*), *t* = 100, with *N* = 2, *n* = 2
Cancellariinae (*t9*), *t* = 0, with *N* = 3, *n* = 0
Plesiotritoninae (*t10*), *t* = 0, with *N* = 1, *n* = 0

The raw data used for these analyses were taken from the species lists shown in Appendix 1. The subprovincial combined index, incorporating the endemicity of these 10 families and subfamilies and defined by the relationship

$$S = \sum_{n=1}^{10} \frac{tn}{10},\ S > 25$$

yields the index quantity *S* = 34.8. This high level of subprovincial endemism (*S* > 25) demonstrates that the Yucatanean Subprovince is a strong, distinct biogeographical unit.

Of particular interest within the Yucatanean Subprovince is the presence of the relictual endemic species *Melongena (Rexmela) bispinosa*, which is a common fossil in the mid-Pleistocene Bermont Formation and late Pleistocene Fort Thompson Formation of southern Florida (Petuch and Roberts, 2007). This small melongenid's presence in the paleontological record of Florida demonstrates that the species was widespread throughout the Gulf of Mexico during the Pleistocene but subsequently has become extinct over most of its range and has managed to survive in the coastal lagoons of the Yucatan Peninsula. Another contemporary late-Pleistocene Floridian species that has managed to survive along the Yucatan Peninsula is the large turbinellid *Turbinella wheeleri*, a common fossil in some facies of the Fort Thompson Formation (Petuch, 1994, 2004; Petuch and Roberts, 2007). This Yucatanean Subprovince endemic has been collected only around Isla Holbox and the Laguna Yalahau, at the extreme northeastern edge of the subprovincial boundaries.

The marginellid *Prunum rostratum*, also found within these lagoon systems, is a prominent fossil in the mid-Pleistocene Bermont Formation of southern Florida (Petuch, 1994) and represents a relictual taxon. The presence of these Pleistocene relics indicates that the entire area is geographically heterochronous and represents a primary relict pocket within the Carolinian Province.

Yucatanean coastal lagoons

The northern and western coastlines of the Yucatan Peninsula, along the States of Campeche, Yucatan, and Quintana Roo, contain a series of mangrove-lined lagoons that harbor low-diversity malacofaunas with high levels of endemicity. Principal among these are the large Laguna del Carmen of Campeche State, the Laguna Celestun of Yucatan State, and the Laguna Yalahau of Quintana Roo State. Unlike the coastal lagoons of the Texan Subprovince, the Yucatanean lagoons are far more stable oceanographically, with relatively constant salinities and temperatures; artifacts of lower latitude and lack of rivers and freshwater effluent in this area. A number of important endemic gastropods inhabit these lagoons, including the following (several illustrated in Figures 5.4 and 5.5):

Modulidae
  Modulus cf. calusa (probably an unnamed species)
Muricidae
  Phyllonotus mexicanus
Melongenidae
  Melongena (Rexmela) bispinosa (and high-spired spineless form *martiniana*)
Busyconidae
  Sinistrofulgur perversum
Turbinellidae
  Turbinella wheeleri (Laguna de Yalahau area only; referred to as *T. scolymoides* by
      Vokes, 1964)
Marginellidae
  Prunum rostratum (also found off Cancun)
Conidae
  Gradiconus maya

The Yucatan coastal areas are still largely unexplored malacologically, and future research will doubtlessly uncover many new relictual taxa and unnamed species.

Endemism on the offshore Yucatan Banks and deep-water areas

The greatest amount of endemism within the Yucatanean Subprovince is found on the offshore carbonate banks that skirt the entire Yucatan Peninsula. The most prominent of these, the Campeche Bank off the States of Campeche and Yucatan, represents the second largest carbonate platform in the Gulf of Mexico, being only slightly smaller than the Florida platform. This shallow carbonate area is dotted with coral reef complexes and small coral cays, the most important being the Alacran Reefs and Cayos Arcas, Cayos Nuevo, and Cayo Arenas. Having remained in eutropical oceanographic conditions throughout the Pleistocene (Petuch, 2004), the entire set of banks and cays has acted as a refugium for many taxa from the Caloosahatchian Paleoprovince. Of special importance on these banks is the genus *Siphovasum*, an elongated turbinellid that is endemic to the area. Some

Figure 5.4 Endemic gastropods of the Yucatanean Subprovince coastal lagoon systems: (A, B) *Gradiconus maya* Petuch and Sargent, 2011, holotype, length 27.9 mm. (C) *Phyllonotus mexicanus* (Petit, 1852), length 78 mm. (D, E) *Melongena (Rexmela) bispinosa* (Philippi, 1844), length 40 mm. (F) *Melongena (Rexmela) bispinosa* form *martiniana* Philippi, 1844, length 46 mm.

of the more important species from these Yucatanean offshore banks include the following (several illustrated in Figures 5.5 and 5.6):

Turritellidae
  Torculoidella lyonsi
  Torculoidella yucatecanum
Personiidae
  Sassia lewisi (also found off Barbados)
Muricidae
  Siratus caudacurta
  Vokesimurex cf. anniae (the true *anniae* is an early Pleistocene fossil from Florida)
  Vokesimurex sallasi
Fasciolariidae
  Cinctura branhamae
Busyconidae
  Busycoarctum coarctatum
  Fulguropsis cf. feldmanni
  Lindafulgur candelabrum
Buccinidae
  Engina dicksoni (see Appendix 2)

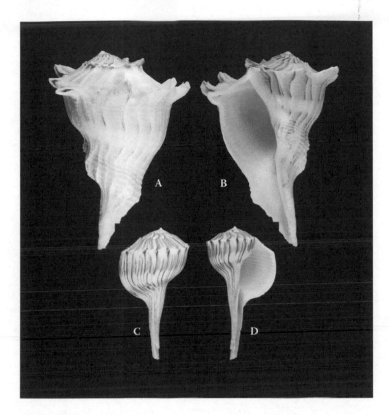

Figure 5.5 Endemic Busyconid gastropods of the Yucatanean Subprovince: (A, B) *Sinistrofulgur perversum* (Linnaeus, 1758), length 226 mm. (C, D) *Busycoarctum coarctatum* (Sowerby, 1825), length 150 mm.

Volutidae
   Scaphella contoyensis
   Scaphella macginnorum
   Scaphella junonia butleri
Turbinellidae
   Siphovasum latiriforme
Marginellidae
   Prunum catochense
   Prunum labiatum
   Prunum lipei
Olividae
   Oliva (Americoliva) contoyensis
   Oliva (Americoliva) maya
Conidae
   Gradiconus sennottorum
   Lindaconus therriaulti (see Appendix 2)
Conilithidae
   Jaspidiconus new species (similar to *pealii* from the Florida Keys)
Terebridae
   Myurellina stegeri

Figure 5.6 Yucatanean Subprovince endemic gastropods from offshore banks: (A) *Scaphella junonia butleri* Clench, 1953, length 114 mm. (B) *Oliva (Americoliva) maya* Petuch and Sargent, 1986, length 60 mm. (C) *Gradiconus sennottorum* (Rehder and Abbott, 1951), holotype, length 35 mm. (D) *Oliva (Americoliva) contoyensis* Petuch and Sargent, 1986, holotype, length 37 mm. (E) *Engina dicksoni* Petuch, new species, holotype, length 18 mm (see Appendix 2). (F) *Lindaconus therriaulti* Petuch, new species, holotype, length 43 mm (see Appendix 2).

These eutropical carbonate-dwelling endemic species occur sympatrically with most of the classic Carolinian index taxa, such as *Macrocypraea (Lorenzicypraea) cervus*, *Hexaplex fulvescens*, *Triplofusus papillosus*, *Dauciconus amphiurgus*, and *Kohniconus delessertii*, demonstrating that the offshore banks fall within the boundaries of the Carolinian Province. Of special interest in the deep-water areas of the Yucatanean Subprovince is the presence of the personiid gastropod *Sassia lewisi*, which has a disjunct, bifurcated range, being found off the Yucatan Peninsula and the Barbados Seamount (Harasewych and Petuch, 1980).

chapter six

Northern subprovinces of the Caribbean Province

The northern subprovinces of the Caribbean Province, the Bermudan, Bahamian, and Antillean, are characterized as being composed primarily of archipelagos of islands. Unlike the malacofaunas of the adjacent Nicaraguan and Venezuelan Subprovinces, which contain long stretches of continental coastlines, the nonvagile (nondispersing) mollusks of the northern subprovinces are highly susceptible to genetic isolation and allopatric speciation. This has led to the evolution of numerous endemic species swarms in the shallow water areas of many of the island chains, a pattern analogous to that seen in the terrestrial environments of the Galapagos Islands. All three biogeographical subdivisions have permanent eutropical conditions and are continuously bathed in warm water, with the Gulf Stream flowing across the Bermudan Subprovince and with the Antillean and Caribbean Currents surrounding the Bahamian and Antillean Subprovinces. Between the Bermuda Islands, the Bahamas, the Greater Antilles Island Arc, and the Belizean Great Barrier Reef and Atolls, over 8,000 islands and carbonate cays are now known from the areas of the three subprovinces. Altogether, this immense archipelago offers a bewildering array of marine habitats and houses some of the most spectacular malacofaunas known from the western Atlantic.

Molluscan biodiversity in the Bermudan Subprovince

At latitude 32.3 degrees north, the Bermudan Islands represent the extreme northern edge of the Caribbean Province and also contain the farthest north zonated coral reefs in the western Atlantic. Although a few hardy hermatypic scleractinian corals do range to a higher latitude on the banks offshore of Cape Hatteras (35.25 degrees north), these do not form the large-scale deterministic carbonate systems seen on Bermuda. The tropical nature of the 122 islands that make up the Bermudan Archipelago was first noted by earlier biogeographers such as Woodward, Warmke and Abbott, and Briggs, and this led to the islands being placed within the Caribbean Province. Being at the edge of the tropical western Atlantic region, the Bermudan molluscan fauna was highly susceptible to fluctuating water temperatures during the late Pleistocene, when the marine climate cooled as much as 6 degrees Celsius below present day water temperatures (Emiliani, 1955). This oscillating temperature regime, along with glacioeustatic sea level fluctuations, placed a tremendous amount of ecological stress upon the Bermudan malacofaunas, leading to increased extirpation and evolution.

Compared with the rest of the Caribbean Province, the Bermudan Subprovince contains an impoverished fauna, with only a few of the widespread Caribbean index taxa having been able to establish themselves that far out into the open Atlantic. Some of these include the strombids *Eustrombus gigas, Lobatus raninus, Aliger gallus,* and *Macrostrombus costatus;* the ovulids *Cyphoma gibbosum* and *Cyphoma signatum;* the cypraeids *Luria cinerea* and *Erosaria acicularis;* the muricid *Phyllonotus pomum;* and the conids *Gladioconus mus* and *Stephanoconus regius*. Besides this complement of classic Caribbean taxa, a number of endemic taxa have evolved on the Bermuda Seamount, and these support the separation of the Bermudan Subprovince from the rest of the Caribbean Province. The data supplied in Appendix 1 demonstrate that this area is a discrete biogeographical entity that can be defined by the percentages of endemism in the 10 key index taxa. Based on the relationship

$$t = \frac{n}{N}(100), \, t > 25$$

these are as follows:

Modulidae ($t1$), $t = 100$, with $N = 1$, $n = 1$
Turbinellidae ($t2$), $t = 0$, with $N = 0$, $n = 0$ (absent from the Bermudan Subprovince)
Conidae ($t3$), $t = 33$, with $N = 3$, $n = 1$
Conilithidae ($t4$), $t = 100$, with $N = 1$, $n = 1$
Muricinae ($t5$), $t = 33$, with $N = 6$, $n = 2$
Fasciolariinae ($t6$), $t = 0$, with $N = 0$, $n = 0$ (absent from the Bermudan Subprovince)
Lyriinae ($t7$), $t = 0$, with $N = 0$, $n = 0$ (absent from the Bermudan Subprovince)
Olivinae ($t8$), $t = 100$, with $N = 1$, $n = 1$
Cancellariinae ($t9$), $t = 0$, with $N = 0$, $n = 0$ (absent from the Bermudan Subprovince)
Plesiotritoninae ($t10$), $t = 0$, with $N = 1$, $n = 0$

The subprovincial combined index, incorporating the endemicity of these 10 families and subfamilies and based on the relationship

$$S = \sum_{n=1}^{10} \frac{tn}{10}, \, S > 25$$

yields the quantity $S = 46.2$. This high level of subprovincial endemism ($S > 25$) demonstrates that the Bermudan Subprovince is a differentiable biogeographical entity. The subprovincial combined index would probably be even higher, approaching provincial status, if there were not so many missing index families. Some of the shallow-water, carbonate-environment, and reef-associated Bermudan endemic gastropods include the following (two shown in Figure 6.1):

Fissurellidae
　Fissurella barbadensis bermudensis
Cerithiidae
　Cerithium bermudae (possibly a subspecies of *lutosum*)
Rissoidae
　Alvania minuscula
Truncatellidae
　Truncatella piratica
Caecidae
　Caecum tornatum
Muricidae
　Dermomurex cf. glicksteini (probably a new species)

Figure 6.1 Endemic gastropods of the Bermudan Subprovince: (A, B) *Bermudaconus lightbourni* (Petuch, 1986), holotype, length 35 mm. (C) *Jaspidiconus mindanus bermudensis* (Clench, 1942), holotype, length 42 mm. (D, E) *Timbellus lightbourni* (Harasewych and Jensen, 1979), holotype, length 39 mm. (F) *Oliva (Americoliva) bifasciata jenseni* Petuch and Sargent, 1986, holotype, length 51 mm.

Olividae
   Oliva (Americoliva) bifasciata jenseni
Conilithidae
   Jaspidiconus mindanus bermudensis
Cimidae
   Graphis lightbourni

The deep-water (200–400 m depths) areas along the edges of the Bermuda Seamount also house a highly endemic malacofauna. For over 100 years, since the beginning of oceanographic research in Bermuda, these deep-water environments were largely unexplored and their levels of molluscan endemism could only be conjectured. This lack of knowledge was reversed in the mid-1970s, when an amateur Bermudan malacologist named John ("Jack") Lightbourn began to sample these bathyal zone environments using deeply submerged lobster traps. These baited traps were left in place for weeks, allowing for the capture of shell-bearing hermit crabs. Using this collecting technique, Lightbourn was able to sample an entire new, previously unexplored Bermudan ecosystem, and many unexpected new deep-water endemic genera and species were collected (e.g., *Bermudaconus lightbourni*; Petuch, 1986a). Some of these included the following (two illustrated in Figure 6.1):

Pleurotomariidae
    Entemnotrochus adansonianus bermudensis
    Perotrochus quoyanus insularis
Muricidae
    Timbellus lightbourni
Fasciolariidae
    Fusinus lightbourni
Conidae
    Bermudaconus lightbourni (see Appendix 2)
    Conasprelloides cf. villepini (small, highly colored form; possibly a new species)

More collecting in the deep-water areas off Bermuda will doubtlessly bring to light many more new and interesting endemic taxa.

Molluscan biodiversity in the Bahamian Subprovince

Comprising over 8,000 separate islands and coral cays, the Bahamian Subprovince (Figure 6.2) includes all of the Bahamas and Turks and Caicos and also the northern coast of Cuba. With the exception of the mountainous coast of mainland Cuba, the Bahamian Subprovince is composed of low, flat islands made up of coralline and oölitic limestones.

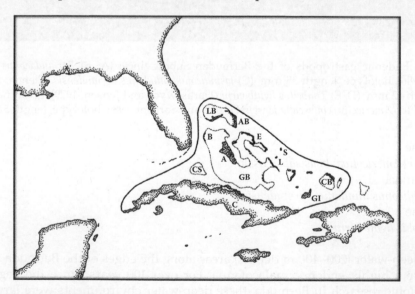

Figure 6.2 Map of the Bahamian Subprovince of the Caribbean Province, showing some of the important geographical features. LB = Little Bahama Bank; AB = Abaco Islands; B = Bimini Chain; E = Eleuthera Island; A = Andros Island; L = Long Cay; GB = Great Bahama Bank; CS = Cay Sal Bank; C = Cuba; S = San Salvador Island; CB = Caicos Bank; GI = Great Inagua Island.

This biogeographical subdivision is centered on the Bahama Banks, which include the Little Bahama Bank in the north, the Great Bahama Bank in the south, and the Cay Sal Bank in the west. These immense carbonate platforms are among the largest in the western Atlantic and house a set of marine environments that are unique in the entire Caribbean area. On these banks, many of the island groups and coral reef complexes are isolated from each other by wide stretches of open, barren oölitic sand sea floors. These submarine deserts are essentially sterile and devoid of nutrients and act as ecological barriers between the productive reef complexes, allowing for the evolution of whole suites of endemic gastropods through allopatric speciation. On some of the larger banks with extensive oölite sand deserts, such as the Great Bahama Bank, as many as five centers of speciation have developed, each with its own endemic species complexes. Some of these include the area around North and South Bimini (Figure 6.3) and the Bimini Chain, the area around Eleuthera Island, the cays near New Providence Island, the Ragged Islands, and the Exuma Cays. A similar pattern is also seen on the Little Bahama Bank, where at least three centers of speciation have developed: the areas around Grand Bahama Island, the Sale Cays, and the Abaco Islands. The coral islands along northern Cuba, the islands of Cay Sal Bank, and the area of the Turks and Caicos are all still virtually unexplored

Figure 6.3 View of the western side of South Bimini Island, Bimini Chain, Bahamas, showing the typical "iron rock" eroded limestone shoreline. The endemic Bimini horn shell, *Cerithium biminiense*, lives in the supratidal pools along this rock shoreline and the Bimini endemic Donna's cone shell, *Purpuriconus donnae*, occurs under coral rubble just a few meters offshore. (From E. Petuch, archival photograph, 1976).

and increased collecting in these areas will undoubtedly bring to light many more new endemic taxa.

Because of its disjunct, patchwork-quilt arrangement of separate malacofaunas, the Bahamian Subprovince is one of the richest in the Caribbean and contains a high level of endemism. The raw data listed in Appendix 1 show that the Bahamian Subprovince is a discrete biogeographical entity and can be defined by the percentages of endemism in the 10 key index taxa. Based on the relationship

$$t = \frac{n}{N}(100), t > 25$$

these are as follows:

Modulidae ($t1$), $t = 50$, with $N = 2$, $n = 1$
Turbinellidae ($t2$), $t = 0$, with $N = 2$, $n = 0$
Conidae ($t3$), $t = 66$, with $N = 30$, $n = 20$
Conilithidae ($t4$), $t = 57$, with $N = 7$, $n = 4$
Muricinae ($t5$), $t = 50$, with $N = 22$, $n = 11$
Fasciolariinae ($t6$), $t = 0$, with $N = 1$, $n = 0$
Lyriinae ($t7$), $t = 100$, with $N = 2$, $n = 2$
Olivinae ($t8$), $t = 33$, with $N = 3$, $n = 1$
Cancellariinae ($t9$), $t = 25$, with $N = 4$, $n = 1$
Plesiotritoninae ($t10$), $t = 33$, with $N = 3$, $n = 1$

The subprovincial combined index, incorporating the endemicity of these 10 families and subfamilies and based on the relationship

$$S = \sum_{n=1}^{10} \frac{tn}{10}, S > 25$$

yields the quantity $S = 53.9$. This high level of endemism, the second largest of any of the Caribbean subprovinces, readily demonstrates that the Bahamian Subprovince is a strong biogeographical entity and that it stands apart from the other adjacent faunal areas.

Endemism on the Bahama Banks

Of the key index taxa reviewed in this book, the Family Conidae is the most prominent and evolutionarily important on the Bahama Banks. Of the 10 western Atlantic conid genera found in the Bahamas, the genus *Purpuriconus* has undergone the largest species radiation, with at least 10 endemic species having evolved on the Little Bahama, Great Bahama, and Cay Sal Banks alone (Petuch, 1992a, 1992b, 1998a, 2000). A smaller radiation of the conid genus *Magelliconus* and one of the conolithid genus *Jaspidiconus* also occur along with the larger *Purpuriconus* species complex. Altogether, these genera, and the conids *Tuckericonus* and *Cariboconus*, have produced the single largest and most spectacular shallow-water cone shell fauna found anywhere in the Caribbean. The described species of Bahamian Banks endemic conids and conilithids include the following (several illustrated in Figures 6.4, 6.5, and 6.6):

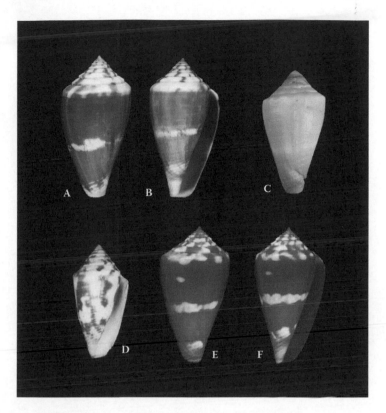

Figure 6.4 Endemic cone shells of the Little Bahama Bank: (A, B) *Purpuriconus jucundus* (Sowerby, 1887), length 34 mm. (Photo by A. Poremski.) (C) *Purpuriconus harasewychi* (Petuch, 1987), holotype, length 26 mm. (D) *Jaspidiconus branhamae* (Clench, 1953), holotype, length 27.5 mm. (E, F) *Purpuriconus lucaya* (Petuch, 2000), length 35 mm. (Photo by A. Poremski.)

Gastropoda

Family Conidae

   Cariboconus sahlbergi (endemic to the Bimini Chain, Great Bahama Bank)

   Magelliconus eleutheraensis (Eleuthera Island; see Appendix 2)

   Magelliconus jacarusoi (endemic to Exuma Sound area, Great Bahama Bank)

   Magelliconus zylmani (endemic to the Bimini Chain and Andros Island)

   Purpuriconus abbotti (endemic to Eleuthera Island, Great Bahama Bank)

   Purpuriconus arangoi (widespread across the Bahamas; also found in Cuba)

   Purpuriconus caysalensis (endemic to Cay Sal Bank)

   Purpuriconus donnae (endemic to the Bimini Chain, Great Bahama Bank)

   Purpuriconus harasewychi (endemic to the Sale Cays, Little Bahama Bank)

   Purpuriconus jucundus (endemic to the Abaco Islands, Little Bahama Bank)

   Purpuriconus lucaya (endemic to Grand Bahama Island, Little Bahama Bank)

   Purpuriconus ortneri (endemic to the New Providence Island area, Great Bahama Bank)

   Purpuriconus richardbinghami (endemic to the Bimini Chain, Great Bahama Bank)

   Purpuriconus stanfieldi (endemic to New Providence and Exuma Sound, Great Bahama Bank)

   Purpuriconus theodorei (endemic to the Providence Island area, Great Bahama Bank)

   Tuckericonus bahamensis (endemic to the Bimini Chain, Great Bahama Bank)

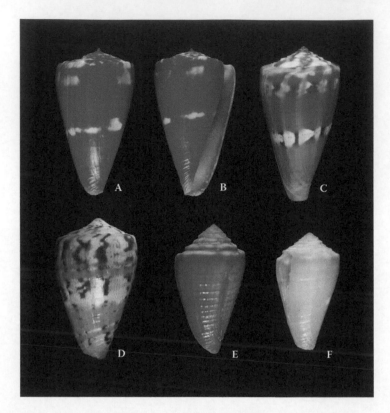

Figure 6.5 Endemic cone shells of the Great Bahama Bank: (A, B) *Purpuriconus richardbinghami* (Petuch, 1992), holotype, length 35 mm. (C) *Purpuriconus theodorei* (Petuch, 2000), length 37 mm. (Photo by A. Poremski.) (D) *Purpuriconus abbotti* (Clench, 1942), holotype, length 35 mm. (E) *Purpuriconus donnae* (Petuch, 1998), holotype, length 28 mm. (F) *Magelliconus zylmanae* (Petuch, 1998), holotype, length 22 mm.

Family Conilithidae
   Jaspidiconus branhamae (endemic to the Abaco Islands, Little Bahama Bank)
   Jaspidiconus exumaensis (Exuma Sound; see Appendix 2)
   Jaspidiconus oleiniki (Bimini Chain; see Appendix 2)

Recently, several more new species of *Jaspidiconus*, *Magelliconus*, and *Purpuriconus* have been discovered on different island groups around the Bahamas, demonstrating that the Bahamian conid fauna may actually be far larger than is presently thought. This rich cone shell fauna coexists with a large number of other distinctive endemic Bahamian Bank gastropods, producing unusual molluscan assemblages unlike any others found in the western Atlantic. Some of the other classic Bahamian Banks endemic gastropods include the following (several shown in Figures 6.4, 6.5, 6.6, 6.7, and 6.8):

Cerithiidae
   Cerithium biminiense
   Fastigiella carinata
Rissoellidae
   Rissoella galba

Figure 6.6 Endemic gastropods of the Bahamas and Turks and Caicos: (A) *Muricopsis zylmanae* Petuch, 1993, length 38 mm. (B) *Enaeta cylleniformis* (Sowerby I, 1844), length 21 mm. (C) *Magelliconus jacarusoi* (Petuch, 1998), length 23 mm. (D) *Purpuriconus stanfieldi* (Petuch, 1998), length 32 mm. (Photo by A. Poremski.) (E) *Polystira bayeri* Petuch, 2001, holotype, length 26 mm. (F) *Purpuriconus liratus* (Reeve, 1844), holotype, length 17.4 mm.

Modulidae
  Modulus cf. modulus
  Modulus honkerorum (see Appendix 2)
Triviidae
  Pusula pacei
Muricidae
  Chicoreus dunni
  Dermomurex binghamae
  Dermomurex worsfoldi
  Murexiella deynzerorum (see Appendix 2)
  Muricopsis sunderlandi
  Muricopsis zylmanae
  Muricopsis honkeri (see Appendix 2)
  Siratus vokesorum
Volutidae
  Enaeta cylleniformis
  Enaeta lindae (see Appendix 2)
Marginellidae
  Prunum pellucidum

Figure 6.7 New discoveries from the Bahama Banks: (A, B) *Modulus honkerorum* Petuch, new species, holotype, width 9 mm. (C, D) *Murexiella deynzerorum* Petuch, new species, holotype, length 21 mm. (E, F) *Muricopsis honkeri* Petuch, new species, holotype, length 21 mm. (G) *Polygona paulae* Petuch, new species, holotype, length 48 mm. The descriptions of these new taxa are given in Appendix 2.

Terebridae
  Strioterebrum biminiensis
  Strioterebrum pacei
Turridae
  Polystira bayeri (lives in shallow depths of 2–3 m)

Many of these species occur throughout the Bahama Banks and farther south on the smaller banks of the Turks and Caicos Islands. Although having essentially a typical Bahamian Subprovince malacofauna, the Turks and Caicos have also evolved a number of endemic species of *Purpuriconus*. Some of these include *Purpuriconus liratus* (Figure 6.6F), *P. dianthus*, and *P. primula*, all of which are known to be closely related to their congeners on the Little and Great Bahama Banks and the Cay Sal Bank. Likewise, *Purpuriconus olgae* and *P. havanensis*, which are endemic to northern Cuba, are also closely related to their Bahamian congeners and together represent the northwestern component of the Bahamian Subprovince conid radiation. The Turks and Caicos and the northern coast of Cuba are among the least studied areas of the Bahamian Subprovince, and intensive field research and collecting will surely yield entire new endemic cone shell faunas.

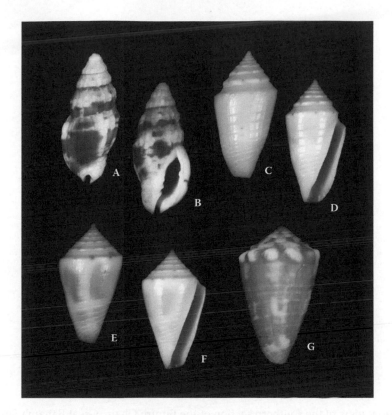

Figure 6.8 New discoveries from the Bahama Banks: (A, B) *Enaeta lindae* Petuch, new species, holotype, length 13 mm. (C, D) *Jaspidiconus exumaensis* Petuch, new species, holotype, length 15.2 mm. (E, F) *Jaspidiconus oleiniki* Petuch, new species, holotype, length 15 mm. (G) *Magelliconus eleutheraensis* Petuch, new species, holotype, length 19 mm. The descriptions of these new taxa are given in Appendix 2.

Endemism in Bahamian deep-water areas

The western edge of the Great Bahama Bank, along the Bimini Chain of islands, contains one of the most unusual deep-water malacofaunas found anywhere in the western Atlantic. This bank margin forms an almost vertical submarine cliff, dropping from depths of 3 m to over 1,000 m less than 1 km from shore (i.e., the Bimini Wall) (Figure 6.9). At a depth of around 400 m, a narrow shelf of only 0.25 km width protrudes from this cliff face and runs the entire length of the island chain (i.e., the Bimini Terrace). This deep-water shelf, interestingly enough, is flooded with warm water of around 16–20 degrees Celsius, the result of a depression in the eastern side of the Gulf Stream. This depressional downwelling of the current is due to a banking effect caused by the sharp northward turn (northward deflection) of the Gulf Stream as it flows into the Straits of Florida (Petuch, 2002; Oleinik, Petuch, and Aley, 2012). While the 400 m depth area along the Bimini Wall has water temperatures of 16–20 degrees Celsius, the same depth on the Florida side of the Strait has much colder water temperatures, averaging only 7.5 degrees Celsius. Because of this water temperature downwelling, the bathyal zone along the Bimini Chain has a unique deep-water tropical malacofauna with a high degree of endemism.

Figure 6.9 Aerial view of the southern section of the Bimini Chain, Great Bahama Bank, Bahamas, showing the steep dropoff along the western side (the Bimini Wall and Bimini Shelf; note the dark-blue color of the water on the lower left). A rich and highly unusual molluscan fauna occurs along the Bimini Shelf at a depth of 400 m, including such rarely seen Bahamian deep-water endemics as *Lindaconus lindae, Persicula bahamsensis, Oliva (Americoliva) bahamasensis, Scaphella (Clenchina) bimini-ensis,* and *Chickcharnea fragilis.* (From E. Petuch, archival photograph, 1977.)

One of the more prominent and abundant gastropods on the Bimini Terrace is the cone shell *Lindaconus lindae* (Figure 6.10D), which sometimes occurs in such high numbers that the accumulation of its dead shells forms a *"Conus* pavement" on the sea floor (Petuch, 2002). This dense aggregation of living and dead cone shells, in turn, acts as a substrate for the attachment of dense thickets of delicate branching stylasterine hydrocorals, stalked crinoid sea lilies, and thin, net-like sponges. While on a research cruise on the Florida Institute of Oceanography (FIO) R/V *Bellows* in 2002, I found that this beautiful and rich warm-water upper bathyal environment supported 24 species of gastropods, with eight new endemic taxa (see Petuch, 2002, for illustrations of this unusual fauna). Subsequent dredging cruises have increased this number to 77 species and have also brought to light several more new species (Oleinik et al., 2012). Some of the more important gastropods that are endemic to the Bimini Shelf and adjacent areas of the eastern Straits of Florida include the following (all 13 species described by Petuch and Sargent, 1986; Petuch, 1987; Petuch, 2002; and Oleinik, Petuch, and Aley, 2012: with several illustrated in Figure 6.10):

Vermetidae
　　Serpulorbis squamolineatus
Turritellidae
　　Vermicularia bathyalis

Figure 6.10 Deep-water endemic Bahamian gastropods: (A) *Persicula bahamasensis* Petuch, 2002, holotype, length 8.5 mm. (B) *Dalliconus pacei* (Petuch, 1987), holotype, length 19 mm. (C) *Conasprelloides leekremeri* (Petuch, 1987), holotype, length 30 mm. (D) *Lindaconus lindae* (Petuch, 1987), holotype, length 31 mm. (E) *Polystira starretti* Petuch, 2002, length 44 mm. (F) *Oliva (Americoliva) bahamasensis* Petuch and Sargent, 1986, length 44 mm.

Buccinidae
    Antillophos bahamasensis
    Antillophos freemani
    Chickcharnea fragilis
Olividae
    Oliva (Americoliva) bahamasensis
Olivellidae
    Olivella (Macgintiella) biminiensis
Marginellidae
    Persicula bahamasensis
Volutidae
    Scaphella (Clenchina) biminiensis
Conidae
    Conasprelloides leekremeri
    Lindaconus lindae
Conilithidae
    Dalliconus pacei
Turridae
    Polystira starretti

Of special interest in the *"Conus* pavement" molluscan community is the genus *Chickcharnea* Petuch, 2002 (the ghost whelk, named for the ghost-like Chickcharnees of Biminian mythology), a shiny, pure white *Ptychosalpinx*-type buccinid that has never been collected anywhere else in the Caribbean area. The ghost whelk and the other endemic Bimini Shelf taxa occur together with a rich and vibrant malacofauna composed of many well-known classic northern Caribbean deep-water species (see Petuch, 2002; Oleinik et al., 2012, for complete species lists and illustrations). Some of the more abundant and characteristic species found on the *Conus* pavement and stylasterine hydrocoral thickets include the following:

Pleurotomariidae
    Entemnotrochus adansonianus
Fissurellidae
    Diodora (Glyphis) tanneri
Xenophoridae
    Tugurium caribaeum
Tonnidae
    Eudolium crosseanum
Ranellidae
    Bursa (Colubrellina) finlayi
    Pisanianura grimaldii
Muricidae
    Poirieria pazi
    Siratus yumurinus
Columbariidae
    Peristarium electra
Fasciolariidae
    Bullockus macmurrayi
    Fusinus halistreptus
Turbinellidae
    Exilia meekiana
Volutidae
    Scaphella (Clenchina) atlantis
    Scaphella (Clenchina) bermudezi
    Scaphella (Clenchina) luizcoutoi (also found off the Turks and Caicos Islands)
Mitridae
    Mitra antillensis
Terebridae
    Myurellina evelynae
Architectonicidae
    Architectonica sunderlandi

The Bahamas also have a large number of other interesting deep-water areas, primary among these being the Tongue-of-the-Ocean, a deep embayment that runs through the middle of the Great Bahama Bank. Like the Bimini Wall, the edges of the Tongue-of-the-Ocean are steep-sided, forming almost sheer vertical underwater cliffs that drop down to over 2,000 m depths. Very little sampling has been done in this enclosed deep canyon, but it is assumed that the malacofauna would be highly endemic. Based upon collections of unnamed giant deep-water turrid gastropods collected by the University of Miami research cruises in the 1960s and 1970s (personal observations), this assumption seems to be valid, and the entire fauna may prove to be of special importance. To date, no comprehensive malacological surveys of this unique oceanographic feature have been undertaken.

Molluscan biodiversity in the Antillean Subprovince

Ranging from the Great Barrier Reef and coral atolls of Belize and eastern Mexico in the west to the Virgin Islands in the east, and encompassing the southern coast of Cuba, Jamaica, Hispaniola, and Puerto Rico, the Antillean Subprovince (Figure 6.11) is, spatially, the largest Caribbean subdivision. It is also the most diverse in habitat types and marine environments, having continental island coastlines with heavy freshwater river effluent and volcanic rock shores, immense contiguous coral reef complexes, and isolated coral atolls and carbonate banks. Like the Bahamian Subprovince, the Antillean contains several different centers of speciation, each with its own set of endemic taxa. The four large continental-type islands, Cuba, Jamaica, Hispaniola (with Haiti and the Dominican Republic), and Puerto Rico, are all separated by wide, deep channels and straits that act as ecological barriers and barriers to dispersal. Because of this, all four of these islands have evolved unique and characteristic molluscan assemblages, often with high degrees of endemism.

As a discrete biogeographical entity, the Antillean Subprovince can be defined by the percentages of endemism in the 10 key index taxa. Using the raw data listed in Appendix 1 and based on the relationship

$$t = \frac{n}{N}(100), \ t > 25$$

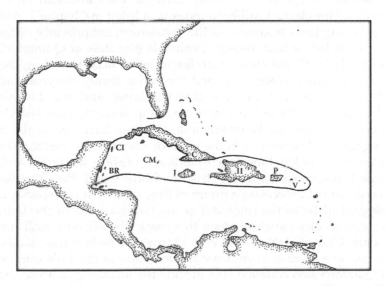

Figure 6.11 Map of the Antillean Subprovince of the Caribbean Province, showing some of the important geographical features. BR = Belizean Great Barrier Reef and Atolls; CI = Cozumel Island; CM = Cayman Islands; J = Jamaica; C = Cuba; H = Hispaniola; P = Puerto Rico; V = Virgin Islands.

these are as follows:

> Modulidae (*t*1), *t* = 0, with *N* = 2, *n* = 0
> Turbinellidae (*t*2), *t* = 0, with *N* = 3, *n* = 0
> Conidae (*t*3), *t* = 50, with *N* = 16, *n* = 8
> Conilithidae (*t*4), *t* = 29, with *N* = 7, *n* = 2
> Muricinae (*t*5), *t* = 22, with *N* = 23, *n* = 5
> Fasciolariinae (*t*6), *t* = 0, with *N* = 1, *n* = 0
> Lyriinae (*t*7), *t* = 75, with *N* = 4, *n* = 3
> Olivinae (*t*8), *t* = 44, with *N* = 9, *n* = 4
> Cancellariinae (*t*9), *t* = 0, with *N* = 4, *n* = 0
> Plesiotritoninae (*t*10), *t* = 50, with *N* = 2, *n* = 1

The subprovincial combined index, incorporating the endemicity of these 10 families and subfamilies and defined by the relationship

$$S = \sum_{n=1}^{10} \frac{tn}{10}, S > 25$$

yields the quantity *S* = 32.4. This high level of subprovincial endemism (*S* > 25) supports the differentiation of the Antillean Subprovince as a separate biogeographical area within the Caribbean Province. As in the case of the Bahamian Subprovince, the Antillean combined index incorporates the endemism of all the highly localized centers of speciation.

Belizean reefs and eastern Yucatan islands

The eastern coast of the Yucatan Peninsula houses a true Caribbean Province malaco-fauna, containing all the classic Caribbean index taxa listed in Chapter 2. Although being part of the same contiguous landmass as the Yucatanean Subprovince of the Carolinian Province, the area of the eastern Yucatan Peninsula (the State of Quintana Roo and the country of Belize) lacks all the classic Carolinian index taxa, including the gastropods *Sinistrofulgur*, *Busycoarctum*, *Lindafulgur* (and the entire family Busyconidae), *Melongena (Rexmela)* (Melongenidae), and *Cinctura* (Fasciolariidae) and the bivalve *Mercenaria* (Veneridae). The ranges of these Carolinian taxa stop abruptly near Isla Contoy on the northeastern tip of the Yucatan Peninsula and, together, form one of the most sharply defined biogeographical boundaries in the western Atlantic. The eastern Yucatan area is characterized by chains of low limestone islands and offshore atoll reefs and also houses the second largest continuous coral reef structure in the world, the Great Barrier Reef of Belize. The conid genus *Cariboconus*, a group of tiny, shallow-water species that occupies the same ecological niche as the intertidal genus *Harmoniconus* in the Indo-Pacific, has undergone a small species radiation along this coast. The offshore atoll reef systems of Belize, particularly Glover's Atoll (one of only 13 true atoll reefs in the Atlantic Ocean), are also known to harbor a number of endemic species. Some of the more interesting of these Belizean and Quintana Roo endemic taxa include the following (with several illustrated in Figures 6.12 and 6.13):

Turbinidae
Bolma (Senobolma) sunderlandi (deep water off the Belizean atolls)

Figure 6.12 Endemic gastropods of Belize and the eastern Mexican islands: (A) *Pterynotus radwini* Harasewych and Jensen, 1979, holotype, length 31 mm. (B, C) *Purpuriconus belizeanus* Petuch and Sargent, 2011, holotype, length 15.56 mm. (D) *Cariboconus kirkandersi* (Petuch, 1987), holotype, length 15 mm. (E) *Cariboconus deynzerorum* (Petuch, 1995), holotype, length 13.5 mm.

Muricidae
   Dermomurex coonsorum (Glover's Atoll; see Appendix 2)
   Pterynotus radwini (deep water off the Belizean atolls)
Olividae
   Oliva (Americoliva) mooreana (deep water off Belizean Atolls; see Appendix 2)
Conidae
   Cariboconus deynzerorum (endemic to Banco Chinchorro Atoll)
   Cariboconus kirkandersi (endemic to Cozumel Island and Isla Mujeres)
   Cariboconus new species (endemic to the Turneffe Islands Atoll)
   Purpuriconus belizeanus (endemic to the reefs of Glover's Atoll)

   The Quintana Roo and Belize coastal areas are still highly unexplored malacologically, and future field research will undoubtedly uncover many more new endemic taxa.

Figure 6.13 New discoveries from the Antillean Subprovince: (A, B) *Planaxis (Supplanaxis) nancyae* Petuch, new species, holotype, length 11 mm. (C, D) *Dermomurex coonsorum* Petuch, new species, holotype, length 10 mm. Note the bright-red encrusting foraminiferan, *Homotrema rubrum*. (E, F) *Oliva (Americoliva) mooreana* Petuch, new species, holotype, length 20 mm. The descriptions of these new taxa are given in Appendix 2.

Endemism in the Greater Antilles

Each of the four main volcanic islands of the Greater Antilles contains a number of distinctive endemic taxa, giving the molluscan assemblages of Cuba, Jamaica, Hispaniola, and Puerto Rico very different appearances. Two groups of gastropods in particular, the conid genus *Purpuriconus* and a species complex centered on the olivid *Oliva (Americoliva) jamaicensis*, have evolved several highly localized endemic taxa, with different closely related species occurring on Jamaica, Hispaniola, and Puerto Rico. Jamaica also houses a number of endemic gastropods, with one of the most noteworthy being the slender, shiny colubrariid *Colubraria sunderlandi*. Some of the more interesting and characteristic Antillean Subprovince endemic gastropods include the following (several illustrated in Figure 6.14):

Figure 6.14 Endemic gastropods of the Greater Antilles: (A) *Oliva (Americoliva) broderipi* Ducros de St. Germaine, 1857, length 26 mm (*O. zombia* Petuch and Sargent, 1986, is a synonym). (B) *Oliva (Americoliva) broderipi* unnamed subspecies, length 25 mm. (C) *Oliva (Americoliva) jamaicensis* Marrat, 1870, length 26 mm. (D) *Jaspidiconus anaglypticus* (Crosse, 1865), holotype, length 17 mm. (E) *Muricopsis warreni* Petuch, 1993, holotype, length 24 mm. (F) *Purpuriconus cardinalis* (Hwass, 1792), length 31 mm.

Cypraeidae
  Macrocypraea (Lorenzicypraea) cervus lindseyi (see Appendix 2)
  Muricidae
  Chicoreus cosmani
  Muricopsis warreni (the *warreni* from Grenada appears to be a new species)
  Pazinotus goesi (endemic to the Virgin Islands)
  Siratus colellai
Colubrariidae
  Colubraria sunderlandi (endemic to Jamaica)
Columbellidae
  Strombina (Cotonopsis) argentea (endemic to Silver Bank off Hispniola)
Volutidae
  Lyria cordis (endemic to Hispaniola)
  Lyria russjenseni (endemic to Puerto Rico)
  Lyria vegai (endemic to Hispaniola)
  Scaphella (Clenchina) neptunia (endemic to Jamaica)

Olividae
  Oliva (Americoliva) broderipi subspecies (endemic to Puerto Rico)
  Oliva (Americoliva) jamaicensis (endemic to Jamaica)
  Oliva (Americoliva) broderipi (endemic to Hispaniola)
Conidae
  Magelliconus cidaris (endemic to Hispaniola)
  Magelliconus explorator (Grand Cayman)
  Magelliconus exquisitus
  Magelliconus sphaecelatus
  Purpuriconus inconstans (= *mayaguensis*?) (endemic to Puerto Rico)
  Purpuriconus cardinalis (endemic to Haiti, especially Gonave Island)
  Purpuriconus speciosissimus (endemic to Mona Island and Puerto Rico)
  Purpuriconus new species (*kulkulcan*-type, endemic to Samana, Dominican Republic)
  Purpuriconus new species (*cardinalis*-type, endemic to Jamaica)
Conilithidae
  Dalliconus lenhilli (endemic to Silver Bank off Hispaniola)
  Jaspidiconus anaglypticus (endemic to Puerto Rico)
  Jaspidiconus mackintoshi (see Appendix 2)
Terebridae
  Strioterebrum juanicum (endemic to Puerto Rico)

Of special interest in the Antillean Subprovince was the recent discovery of a new member of the subgenus *Lorenzicypraea* (Cypraeidae) from the southern coast of Cuba. This group of cowries, represented by *Macrocypraea (Lorenzicypraea) cervus* and its fossil relatives, was previously known only from the Carolinian Subprovince and a few areas along northern Cuba. The new isolated Antillean subspecies, *Macrocypraea (Lorenzicypraea) cervus lindseyi* (Figure 6.15A,B, and C), was discovered in 2001 by renowned diver and collector Glenn Duffy of Sarasota, Florida, but was misidentified as *Macrocypraea cervus peilei* (a Pleistocene fossil known only from Bermuda; see Appendix 2 for the description and discussion). The islands of the Antillean Subprovince, particularly Cuba and Hispaniola, are still relatively unstudied and unsurveyed. Considering the large number of new endemic taxa that have recently been discovered, future field research will definitely bring to light many more interesting discoveries.

Figure 6.15 New discoveries from the Antillean Subprovince: (A, B) *Macrocypraea (Lorenzicypraea) cervus lindseyi* Petuch, new subspecies, holotype, length 53 mm. (This subspecies has been incorrectly referred to as *Macrocypraea cervus peilei*, which is actually an early Pleistocene fossil from Bermuda. Note the distinctive bright-white anterior and posterior extremities of the new subspecies.) (C) *Macrocypraea (Lorenzicypraea) cervus lindseyi* Petuch, new subspecies, paratype, length 50 mm. (D, E) *Jaspidiconus mackintoshi* Petuch, new species, holotype, length 13.7 mm. The descriptions of these new taxa are given in Appendix 2.

chapter seven

Molluscan biodiversity in the Nicaraguan Subprovince

Extending from the Belize–Guatemala Border on the Golfo de Honduras to the Panamá–Colombia Border near the Golfo de Urabá, the Nicaraguan Subprovince encompasses the entire eastern Central American coastline and the adjacent offshore banks and islands (Figure 7.1). This large biogeographical subdivision contains a wide variety of marine environments, including extensive organic-rich muddy coastlines, dense mangrove jungles and brackish coastal lagoons; large offshore carbonate banks and coral reef complexes; and high volcanic islands surrounded by fringing coral reefs. Altogether, these habitats house one of the richest malacofaunas known from the tropical western Atlantic region. Of particular interest within the Nicaraguan Subprovince are several centers of localized speciation, all of which contain highly endemic faunas. The most important of these are as follows: (1) the muddy shoreline of the Miskito Coast; (2) the carbonate banks off Honduras and Nicaragua; (3) the Bay Islands of Honduras (Roatan, Utila, and Guanaja); and (4) the San Blas Islands of Panamá. Using the raw data listed in Appendix 1 and based on the percentages of endemism of the 10 key index taxa found in the four centers of speciation and elsewhere along the Central American coast, the Nicaraguan subprovincial boundaries are readily distinguishable. Applying the relationship

$$t = \frac{n}{N}(100), \ t > 25$$

these are as follows:

Modulidae ($t1$), $t = 33$, with $N = 3$, $n = 1$
Turbinellidae ($t2$), $t = 0$, with $N = 2$, $n = 0$
Conidae ($t3$), $t = 79$, with $N = 34$, $n = 27$
Conilithidae ($t4$), $t = 25$, with $N = 8$, $n = 2$
Muricinae ($t5$), $t = 33$, with $N = 21$, $n = 7$
Fasciolariinae ($t6$), $t = 0$, with $N = 2$, $n = 0$
Lyriinae ($t7$), $t = 100$, with $N = 1$, $n = 1$
Olivinae ($t8$), $t = 20$, with $N = 5$, $n = 1$
Cancellariinae ($t9$), $t = 25$, with $N = 4$, $n = 1$
Plesiotritoninae ($t10$), $t = 0$, with $N = 1$, $n = 0$

The subprovincial combined index, incorporating the endemicity of these 10 families and subfamilies and defined by the relationship

$$S = \sum_{n=1}^{10} \frac{tn}{10}, \ S > 25$$

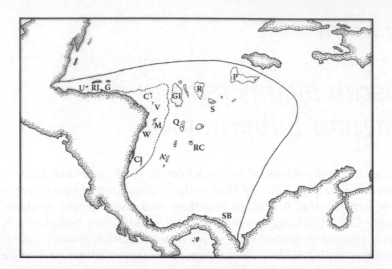

Figure 7.1 Map of the Nicaraguan Subprovince of the Caribbean Province, showing some of the important geographical features. U = Utila Island; RI = Roatan Island; GI = Guanaja Island (The Bay Islands); C = Caratasca Cays; V = Vivorillos Cays; G = Gorda Bank; R = Rosalind Bank; S = Serranilla Bank; P = Pedro Bank; M = Miskito Cays; W = Wawa River Mouth; Q = Quitasueño Bank; RC = Roncador Bank; A = Albuquerque Cays; CI = Corn Island; SB = San Blas Islands.

yields the quantity $S = 31.6$. This high level of subprovincial endemism ($S > 25$) demonstrates that the Nicaraguan Subprovince is a discrete biogeographical subdivision of the Caribbean Province.

The single most prominent and characteristic group of macrogastropods found within the Nicaraguan Subprovince is the genus *Voluta* (Volutidae-Volutinae), which has evolved an endemic radiation of at least 10 species on the isolated offshore banks and muddy coastal areas (several illustrated in Figures 7.2 and 7.3). Since these volutes have very large protoconchs and direct development, nonplanktonic larvae, they are easily trapped and isolated on offshore banks and seamounts and have produced a swarm of sibling species since the late Pliocene (Petuch, 1981b). During the heavy glaciations of the Gelasian Age of the earliest Pleistocene, lowered sea levels (as much as 250 m) allowed these shallow-water volutes to spread across the wide continental shelf and colonize previously unreachable offshore areas. During warm interglacial times, particularly during the Sangamonian Stage of the late Pleistocene, sea levels rose to their present height or higher (20 m or more in the Sangamonian) and flooded the wide shelf with much deeper water. Many of the shallow-water volutes became isolated on these offshore banks and island groups and quickly evolved a suite of closely related allopatric siblings. This unique species radiation, endemic to the Nicaraguan Subprovince, contains the following species:

Family Volutidae
Subfamily Volutinae
    Voluta demarcoi (coastal areas from Punta Patuca and Cabo Camaron to Puerto Cabezas)
    Voluta ernesti (coastal areas from Punta Gorda to the San Blas Islands, Panamá)
    Voluta hilli (endemic to the Gorda Bank and area of the Cajones and Bercero Cays)
    Voluta kotorai (Corn and San Andres Islands, and Gorda, Roncador, Quitasueño, and
        Serrana Banks)
    Voluta lindae (endemic to the San Blas Islands area of Panamá)
    Voluta morrisoni (endemic to Rosalind and Serranilla Banks)

Figure 7.2 Endemic volutes of the Nicaraguan Subprovince: (A) *Voluta kotorai* Petuch, 1981, length 85 mm. (B) *Voluta polypleura* Crosse, 1876, Neotype (USNM 894291), length 46.4 mm. (C) *Voluta demarcoi* Olsson, 1965, length 62 mm (a valid species, different from *V. polypleura*). (D) *Voluta hilli* Petuch, 1987, length 72 mm. (E) *Voluta morrisoni* Petuch, 1980, length 45 mm. (F) *Voluta retemirabila* Petuch, 1981, length 77 mm.

>*Voluta polypleura* (endemic to the Caratasca and Vivorillo Cays)
>*Voluta retemirabila* (endemic to Misteriosa Bank)
>*Voluta sunderlandi* (endemic to Utila Island)
>*Voluta new species* (endemic to the Miskito Cays)

In an ongoing study of the genus *Voluta*, Anton Oleinik (Florida Atlantic University) and I recently discovered that the name *Voluta polypleura*, which Poppe and Goto (1992) applied to the shell previously known as *Voluta demarcoi*, was actually misapplied and incorrect. The real *Voluta polypleura* is a much smaller and more compact shell and, as pointed out by the species' author (Crosse, 1876), completely lacks any "music lines or musical staff pattern" on the body whorl. The larger, more rectangular *Voluta demarcoi* (Figure 7.2C) always has strong color lines encircling the shell midbody, and its general appearance certainly does not match up with Crosse's original description. Unfortunately,

Figure 7.3 Close-up of two live specimens of *Voluta sunderlandi* Petuch, 1987 from Utila Island, Bay Islands, Honduras. The larger specimen with the extended proboscis is 48 mm in length. (Photograph by Kevan Sunderland, 1985.)

Crosse's type specimen of *Voluta polypleura* was lost and could not be located in any museum in Europe. This apparent loss, however, was reversed by the rediscovery of the type locality of the true *Voluta polypleura* and by the collection of several live specimens. One of these so closely matched the illustration of Crosse's species in color, shape, and size that it was decided to make this individual the neotype, thus preserving Crosse's name and removing *demarcoi* from synonymy with *polypleura*.

The true *Voluta polypleura* (Figure 7.2B), with its neotype specimen in the Division of Mollusks at the National Museum of Natural History at the Smithsonian Institution (*Voluta polypleura*, Neotype Number USNM 894291), is now known to be an inhabitant of shallow-water coral reefs around the Caratasca Cays (new type locality) and Vivorillos Cays of Honduras. The common and well-known *Voluta demarcoi* should now be removed from synonymy with *polypleura* and should be considered a full, valid species. Specimens of other western Caribbean *Voluta* species, such as *hilli, kotorai, ernesti,* and *retemirabila,* have been misidentified by many volute workers and shell dealers as subspecies or varieties of *Voluta polypleura* (as incorrectly defined by Poppe and Goto, 1992). These Nicaraguan Subprovince endemic volutes should now be considered to be separate full species, with most having restricted ranges on offshore island chains and isolated banks. Recent collecting on other island groups, such as the Miskito Cays, has uncovered at least three more geographically restricted unnamed species, and more research on these offshore archipelagos will undoubtedly uncover more new taxa.

Coastal Central America

The largest single stretch of muddy shorelines and mangrove jungles found anywhere in the Caribbean area is located along the coasts of Honduras and Nicaragua. These are particularly well developed along the Honduran coast, from Cabo Camaron to Cabo Gracias a Dios and including the giant Laguna de Caratasca, and along the Nicaraguan coast from Laguna Bilwaskarama south to Laguna de Perlas and Bluefields Lagoon. Here, an immense amount of freshwater effluent, filled with clay and mud, pours out into the Caribbean from hundreds of small rivers and creates a continuous flow of fine particulate sediments. Because of the suspended clay and mud, the coastal waters of the Miskito Coast (named for the Miskito Tribe) are perpetually murky and resemble the muddy water seen in large rivers. This muddy-substrate, murky-water environment is rich in nutrients and is highly productive, supporting huge populations of shrimp and scallops and forming the foundation of a major commercial fishing industry.

I had the opportunity to conduct the first comprehensive surveys of the molluscan fauna of the Miskito Coast in 1991 and 1992, under the auspices of the Caribbean Conservation Corporation (CCC) and IRENA (the Nicaraguan governmental environmental agency). Working along the area of Puerto Cabezas and the Wawa River, I discovered a new type of muddy Caribbean environment, complete with several interesting new endemic mollusks (including *Ficus villai, Cancellaria mediamericana, Gradiconus paschalli, Sheldonella lindae, Plicatula miskito, Mactra inceri,* and *Micromactra miskito*; Petuch, 1998b). Interestingly enough, the molluscan fauna of the Miskito Coast more closely resembled that found in the muddy environments of the Panamic Province (western Central America) than it did molluscan faunas found in other muddy areas of the Caribbean Sea. A strong Panamic faunal influence is present in the Miskito molluscan assemblages, with many previously unknown Panamic–Caribbean analogue species pairs. Some of these endemic muddy coastal species include (several illustrated in Figures 7.4 and 7.5):

Gastropoda
Strombidae
  Strombus pugilis nicaraguensis (Caribbean analog of the Panamic *S. gracilior*)
Ficidae
  Ficus villai (endemic to Honduras and Nicaragua)
Muricidae
  Chicoreus bullisi
  Vokesimurex garciui
  Vokesimurex samui
Volutidae
  Voluta demarcoi
  Voluta virescens fulva
Olividae
  Agaronia hilli (Caribbean analog of the Panamic *A. propatula*)
  Agaronia leonardhilli (Caribbean analog of the Panamic *A. testacea*)
Cancellariidae
  Cancellaria mediamericana (Caribbean analog of the Panamic *C. obesa*)
Conidae
  Gradiconus aureopunctatus
  Gradiconus paschalli (Caribbean analog of the Panamic *G. regularis*)
  Lindaconus lorenzianus
Turridae
  Hindsiclava tippetti (Caribbean analog of the Panamic *H. militaris*)
Bivalvia
Noetiidae
  Sheldonella lindae (Caribbean analog of the Panamic *S. olssoni*)

Figure 7.4 Endemic gastropods from the muddy coastlines of the Nicaraguan Subprovince: (A) *Agaronia hilli* Petuch, 1987, length 28 mm. (B) *Agaronia leonardhilli* Petuch, 1987, length 44 mm. (C) *Vokesimurex samui* (Petuch, 1987), length 78 mm. (D) *Vokesimurex garciai* (Petuch, 1987), length 71 mm. (E) *Gradiconus paschalli* (Petuch, 1998), length 24 mm. (F) *Ficus villai* Petuch, 1998, length 67 mm.

Plicatulidae
   Plicatula miskito (Caribbean analog of the Panamic *P. penicillata*)
Mactridae
   Mactra inceri
   Micromactra miskito (Caribbean analog of the Panamic *M. californica*)

In 1905, the Reverend W. H. Fluck, a Moravian missionary to the Miskito Indians and an amateur naturalist, published the results of his shell collecting in the vicinity of the Wawa River mouth. Although meant to be only a cursory overview of the local malacofauna, Fluck's paper (published in the *Nautilus*) did give the description of one new subspecies, the dwarf fighting conch *Strombus pugilis nicaraguensis* (Figure 7.6E,F). This unusual, heavily sculptured strombid appears to be endemic to the Miskito Coast area, where it is particularly common in the region of the Wawa River (the type locality). Small *Strombus pugilis pugilis* from elsewhere in the southern Caribbean have been confused with this endemic Nicaraguan subspecies and are frequently misidentified. The true *nicaraguensis* lacks large spines on the spire whorls and has a shoulder spine arrangement similar to that of the Carolinian *Strombus alatus*. *Strombus pugilis pugilis*, regardless of size, always has a row of very large spines on the penultimate whorl. Characteristically, these spines are always

Figure 7.5 Endemic mollusks from the coastal areas of the Nicaraguan Subprovince: (A) *Pleioptygma helenae* (Radwin and Bibbey, 1972), length 98 mm. (B) *Micromactra miskito* Petuch, 1998, length 30 mm. (C) *Sheldonella lindae* (Petuch, 1998), length 26 mm. (D) *Hindsiclava tippetti* Petuch, 1987, length 69 mm. (E) *Gradiconus aureopunctatus* (Petuch, 1987), holotype, length 17 mm (originally incorrectly stated as coming from the Paraguaná Peninsula of Venzuela but now known to be from the muddy coastlines of Nicaragua and Honduras; the type locality is here emended to "off Puerto Cortez, Honduras, on mud bottom, 3 m depth"). (F) *Chicoreus bullisi* E. Vokes, 1974, length 65 mm.

larger than those on the earlier whorls, contrasting greatly with the small regular spines on the spire of *Strombus pugilis nicaraguensis*. Future DNA work may prove that this small subspecies is genetically separable and may represent a living descendant of one of the small late Pleistocene strombids from Florida, such as *Strombus lindae* and *Strombus dowlingorum* (Petuch and Drolshagen, 2011).

As discussed in Chapter 1, the coastal areas of Honduras and Nicaragua are known to be geographically heterochronous, housing a large component of relictual taxa. These relicts are remnants of the Caloosahatchian Paleoprovince of the Pliocene and early Pleistocene, and they most probably represent surviving populations from the farthest southern Caloosahatchian malacofaunas. While the rest of the paleoprovince became too cold during the mid-Pleistocene climatic degeneration (Illinoian glacial stage), resulting in the extinction of the classic Caloosahatchian taxa, a few remnants found a refugium in the warmer areas at the extreme southern edge of the provincial boundaries. Primary among these Nicaraguan Subprovince Caloosahatchian relicts are the giant miter *Pleioptygma helenae* (Pleioptygmatidae; related to the Caloosahatchee Formation *Pleioptygma lineolata*; Figure 7.5A), *Cerithioclava garciai* (Cerithiidae; related to the Caloosahatchee Formation

Figure 7.6 Gastropods of the Nicaraguan Subprovince: (A) *Knefastia hilli* Petuch, 1990, length 53 mm. (B) *Attenuiconus eversoni* (Petuch, 1987), holotype, length 18 mm. (C) *Cerithioclava garciai* Houbrick, 1985, length 65 mm. (D) *Enaeta reevei* Dall, 1907, length 15 mm. (E, F) *Strombus pugilis nicaraguensis* Fluck, 1905, length 58 mm. (G) *Jaspidiconus sargenti* Petuch, new species, holotype; length 21.7 mm (see Appendix 2).

Cerithioclava caloosaensis; Figure 7.6C), and the unusual giant turrid *Knefastia hilli* (Turridae; closely related to the Caloosahatchee Formation *Knefastia lindae*; Figure 7.6A). These taxa demonstrate that the Honduran–Nicaraguan coastal area qualifies as a secondary relict pocket.

Endemism on the Bay Islands of Honduras

The islands of Roatan, Utila, and Guanaja, collectively referred to as the Bay Islands (Islas de la Bahia), are high, volcanic peaks that rise abruptly from the deep-water areas of the Golfo de Honduras. In being surrounded by deep water and being located as much as 50–100 km offshore, the mollusks of the Bay Islands, particularly the low-dispersibility and nonvagile types, are particularly susceptible to genetic isolate on and allopatric speciation. The islands are fringed by extensive coral reef complexes and clear-water conditions, the complete opposite of the environments along the adjacent muddy coastlines. These reef systems harbor a large component of endemic gastropods, some of which include the following (illustrated in Figure 7.7):

Figure 7.7 Endemic gastropods from the Bay Islands of Honduras. (A) *Jaspidiconus roatanensis* Petuch and Sargent, 2011, length 13.9 mm. (B) *Purpuriconus kulkulcan* (Petuch, 1980), holotype, length 21 mm. (C) *Lindaconus sunderlandi* (Petuch, 1987), holotype, length 33 mm. (D) *Voluta sunderlandi* Petuch, 1987, length 53 mm. (E) *Tenorioconus harlandi* (Petuch, 1987), holotype, length 33 mm. (F) *Dibaphimitra janetae* Petuch, 1987, length 45 mm.

Modulidae
    Modulus hennequini (see Appendix 2)
Columbellidae
    Zaphrona sunderlandi
Fasciolariidae
    Polygona abbotti
    Polygona bessei (see Appendix 2)
    Polygona martini
Volutidae
    Voluta sunderlandi
Mitridae
    Dibaphimitra janetae (may also be found in southern Cuba)
Olividae
    Ancilla (Amalda) pacei
Conidae
    Attenuiconus eversoni
    Cariboconus kalafuti
    Cariboconus magnottei
    Lindaconus sunderlandi (also present on the adjacent Honduran mainland)

> *Purpuriconus kulkulcan* (also present on the adjacent Honduran mainland)
> *Tenorioconus harlandi* (also found on a small stretch of the adjacent Honduran mainland)
Conilithidae
> *Jaspidiconus allamandi* (see Appendix 2)
> *Jaspidiconus sargenti* (see Appendix 2)
> *Jaspidiconus roatanensis*

Several normally rare Caribbean gastropods have been found to be relatively common on the reefs of the Bay Islands, buried deeply in dead coral rubble. Principal among these is the bright-red Glory-of-the-Atlantic cone, *Atlanticonus granulatus*, which occurs together with a bright-red color form of *Purpuriconus kulkulcan*, the red miter *Vexillum histrio*, and a red form of the muricid *Muricopsis deformis*, forming a "red community" deep within the reef rubble zone. These deeply buried, interstitial red gastropods also coexist with red amphinomid polychaetes, red sponges, and red ophiuroid brittle stars.

Honduran and Nicaraguan offshore banks

Far offshore on the wide, shallow continental shelf of the Honduran–Nicaraguan border area, the influence of the muddy coastal waters diminishes and the clean, warm water encourages lush coral growth. Here, in the middle of the carbonate platform, over 20 archipelagos of coral cays have formed, with the largest and most important being the Caratasca, Vivorillos, Bercerro, Cajones, Medialuna, Man-of-War, and Miskito Cays. Together, over 1,000 small coral islands exist among these vast reef complexes, making this one of the four largest shallow-water carbonate areas in the western Atlantic. Farther offshore of the Honduran–Nicaraguan platform, in the central Caribbean Basin, 12 separate smaller carbonate banks have formed on top of volcanic seamounts. These create a stepping stone-like bridge between Jamaica and the Greater Antilles and the mainland of Central America, with the largest and most important being the Gorda, Rosalind, Pedro, Serranilla, Quitasueño, Albuquerque, and Roncador Banks (Figure 7.1). Being surrounded by deep water, the nonvagile benthonic organisms of these banks are prone to genetic isolation and allopatric speciation, and this has led to the evolution of several important endemic gastropods (Petuch, 1980a, 1995a). Some of these include the following (several illustrated in Figures 7.8 and 7.9):

Volutidae
> *Enaeta bessei* (endemic to Rosalind Bank; see Appendix 2)
> *Voluta morrisoni* (endemic to Rosalind and Serranilla Banks)
Conidae
> *Atlanticonus ritae* (endemic to Gorda Bank)
> *Attenuiconus ignotus*
> *Cariboconus bessei* (Caratasca and Vivorillos Cays)
> *Cariboconus flammeacolor* (Rosalind, Gorda, and Serranilla Banks)
> *Purpuriconus rosalindensis* (Rosalind Bank)
> *Tenorioconus julieandrieae* (endemic to Corn Island)
> *Tuckericonus ceruttii*

The Honduras–Nicaragua carbonate platform and the coastal areas of Costa Rica and Panamá also house a large radiation of the conid genus *Gradiconus*, with at least six species. Some of these include (illustrated in Figures 7.8 and 7.9):

Figure 7.8 Endemic cone shells of the Nicaraguan Subprovince: (A) *Tenorioconus panamicus* (Petuch, 1990), holotype, length 24 mm. (B) *Gradiconus rosemaryae* (Petuch, 1990), holotype, length 25 mm. (C) *Atlanticonus ritae* (Petuch, 1995), holotype, length 27.5 mm. (D) *Purpuriconus hilli* (Petuch, 1990), holotype, length 21 mm. (E) *Gradiconus portobeloensis* (Petuch, 1990), holotype, length 31 mm. (F) *Gradiconus ernesti* (Petuch, 1990), holotype, length 29 mm.

> *Gradiconus bayeri*
> *Gradiconus ernesti*
> *Gradiconus garciai*
> *Gradiconus ostrinus*
> *Gradiconus portobeloensis*
> *Gradiconus rosemaryae*

All along this area, the *Gradiconus* species occur sympatrically with the coastal volutids *Enaeta reevei* (Figure 7.6D), *Voluta hilli*, *Voluta demarcoi*, *Voluta ernesti*, and *Voluta virescens* and the muricid *Chicoreus emilyae*, forming an unusual and characteristic molluscan assemblage.

San Blas Archipelago

The San Blas Islands of Panamá form the extreme southeastern edge of the Nicaraguan Subprovince and contain several marine environments that are unique within the Caribbean Province. One of the most interesting and important of these is a substrate type made up of massive coralline algal "reefs" (fused rhodoliths) composed of *Porolithon* and *Lithothamnion* red calcareous algae. Some of these algal bioherms, particularly around

Figure 7.9 Endemic cone shells of the Nicaraguan Subprovince: (A) *Atlanticonus glenni* (Petuch, 1993), length 24 mm (endemic to the San Blas Islands). (B) *Cariboconus flammeacolor* (Petuch, 1992), holotype, length 10.5 mm. (C) *Purpuriconus rosalindensis* (Petuch, 1998), holotype, length 19 mm. (D) *Cariboconus magnottei* (Petuch, 1987), holotype, length 14 mm. (E) *Gradiconus bayeri* (Petuch, 1987), holotype, length 16 mm. (F) *Cariboconus bessei* (Petuch, 1992), holotype, length 15 mm. (G) *Lindaconus lorenzianus* (Dillwyn, 1817), length 59 mm (shell covered with thick raised cords).

Moro Tupo Island, form definite ridges and platforms that act as barriers to wave action (Vermeij, 1978: 88–89).

Several sessile vermetid gastropods, including *Stephopoma myrakeenae* and *Petaloconchus nigricans*, live embedded within the algal rock, forming an unusual molluscan assemblage along with the San Blas endemic cerithiid *Cerithium caribbaeum*. The shallow coral reef areas around the islands also contain large growths of coralline algae that have been found to support a large number of other endemic mollusks. Referred to as the Blasian Subregion (Petuch, 1991), the San Blas Islands and adjacent areas in the Golfo de San Blas represent an "island" of clean carbonate environments in an area of otherwise muddy substrates. With the organic-rich mudflats and mangrove jungles of the Golfo de Urabá to the east and with the open muddy coastline of Panamá and Costa Rica to the west, the Blasian Subregion coral environments are isolated from the other reef systems in the southern Caribbean and have evolved their own distinctive malacofaunas (see Petuch, 1991, 1993a for details on the Blasian ecosystems and for illustrations of many of the rarer species). Some of the more important San Blas Islands and Blasian Subregion endemic gastropods include the following (several illustrated in Figure 7.10):

Figure 7.10 Endemic gastropods of the San Blas Islands: (A) *Atlanticonus cuna* (Petuch, 1998), length 22 mm. (Photo by A. Poremski.) (B) *Voluta lindae* Petuch, 1997, length 33 mm. (C) *Cariboconus brunneofilaris* (Petuch, 1990), holotype, length 14 mm. (D) *Polygona cuna* (Petuch, 1990), length 45 mm. (E) *Prunum leonardhilli* Petuch, 1990, length 19 mm. (F) *Oliva (Americoliva) reticularis ernesti* Petuch, 1990, length 35 mm (may also occur along the coastlines of Panama and Costa Rica).

Liotiidae
 Arene bitleri
Cerithiidae
 Cerithium caribbaeum
Turritellidae
 Torcula marianopsis
Muricidae
 Chicoreus hilli
 Dermomurex cuna (often referred to as *antecessor,* which is a fossil species)
 Murexiella edwardpauli
 Vokesimurex rubidus panamicus (probably a full species)
Fasciolariidae
 Polygona cuna
Columbellidae
 Nassarina dubia
Olividae
 Oliva (Americoliva) reticularis ernesti

Olivellidae
  Olivella (Niteoliva) marmosa (also found in the Bocas Islands, Bocas del Toro)
Marginellidae
  Prunum leonardhilli
Mitridae
  Mitra (Nebularia) leonardi
Volutidae
  Voluta lindae (lives in clean carbonate sand on coral reefs)
Conidae
  Atlanticonus cuna
  Atlanticonus glenni
  Cariboconus brunneofilaris
  Poremskiconus edwardpauli
  Purpuriconus hilli
  Tenorioconus panamicus
Turridae
  Fusiturricula sunderlandi

Many of the deeper water, offshore species were originally collected by James ("Jimmy") Ernest of Balboa, Panamá, who spent many years dredging the areas off Portobelo and Moro Tupo. He is personally responsible for discovering over 20 new species and, through his collecting efforts, has made a tremendous contribution to our knowledge of the molluscan faunas of Caribbean Panamá. Along with James Ernest, two renowned French collectors, Francis Hennequin and Bruno Besse, have also recently discovered many new species from the Caribbean coast of Central America. While diving on the unexplored coral reefs around the Bay Islands of Honduras and on the central Caribbean banks, Besse and Hennequin have uncovered some spectacular new triviids, volutes, cone shells, fasciolariids, and other gastropods, several of which are illustrated here on Figure 7.11. These intrepid French divers have also rediscovered the habitats of many of the rarest and most poorly-known western Caribbean volutes, including the extremely beautiful Kotora's Volute, *Voluta kotorai* (Figure 7.12).

Figure 7.11 New discoveries from the Nicaraguan Subprovince: (A) *Modulus hennequini* Petuch, new species, holotype, width 12 mm. (B, C) *Pusula bessei* Petuch, new species, holotype, length 18 mm. (D, E) *Enaeta bessei* Petuch, new species, holotype, length 10 mm. (F) *Jaspidiconus allamandi* Petuch, new species, holotype, length 16.2 mm. (G) *Polygona bessei* Petuch, new species, holotype, length 33 mm. The descriptions of these new taxa are given in Appendix 2.

Figure 7.12 Kotora's volute, *Voluta kotorai* Petuch, 1981, which is endemic to the Nicaraguan Subprovince, is one of the most beautiful gastropods from the Caribbean Province. This 94 mm specimen was collected in 20 m of water by a commercial lobster diver on Gorda Bank, central Caribbean Sea.

chapter eight

Molluscan biodiversity in the Venezuelan Subprovince

Extending from the Golfo de Urabá in the west to the western coast of Trinidad in the east, and encompassing the entire northern coast of Colombia and the mainland of Venezuela (Figure 8.1), the Venezuelan Subprovince contains one of the top three richest malacofaunas found in the western Atlantic, with over 1,500 species known from the Colombian coast alone (Daccarett and Bossio, 2011). The species richness of the Venezuelan molluscan assemblages is due primarily to the high productivity and high nutrient levels of the local coastal waters, the result of both heavy river effluent and perpetual wind-driven upwelling systems (Petuch, 1982, 1987, 1988; Vermeij and Petuch, 1986). These productive marine environments also support a large commercial fishing industry, primarily for shrimp, and the shrimp boats are often the sole source for collections of deeper water, offshore mollusks. I had the opportunity to work on several Colombian shrimp boats (Vikingos de Colombia Shrimp Company) during the years 1976–1980 and was able to visit and collect in areas that were previously unexplored, primarily along the Goajira Peninsula, off the mouth of the Magdalena River, and in the Golfo de Morrosquillo. Since the shrimpers drag their nets for over three hours at a time, large swaths of sea floor could be sampled, and I collected over 40 new species of macrogastropods from the large piles of marine life that were deposited on deck (Figure 8.2). From these collections, it was apparent that the Venezuelan Subprovince molluscan fauna contains a large component of relict taxa and a very high level of endemicity (Petuch, 1981a).

The subprovince can be subdivided into four main sections, each dominated by a particular sea floor type: (1) the coastline from the Golfo de Urabá to the Magdalena River mouth, which is dominated by muddy, organic-rich bottoms; (2) the Goajira Peninsula and Monjes Islands, which are dominated by shell hash carbonate sea floors and immense sponge bioherms; (3) the Golfo de Venezuela and the Paraguaná Peninsula, which are dominated by clay and diatomaceous oozes, caused by upwellings of cool water; and (4) the coastline from the Golfo Triste to Margarita Island and Trinidad, which is composed of terriginous sediments and gravels, intermixed with fine particulate carbonates and scattered coral reefs. Although some taxa are eurytopic and can exist on all of these substrate types, most are stenotopic and are confined to one or two of the main sea floor types. A classic example of this ecological exclusivity caused by substrate type is seen in the Volutidae, where *Voluta musica* thrives on the clean carbonate sand, shell hash, and clean coarse quartz sand of the Colombian coast from Santa Marta, along the Goajira Peninsula, and into the Golfo de Venezuela. The related *Voluta virescens*, on the other hand, prefers the organic-rich muddy sea floors from Santa Marta westward to the Golfo de Urabá. These two ecologically exclusive volutes occur together in only one small area near Santa Marta, where there is a mixture of carbonates, quartz sand, and organic-rich mud.

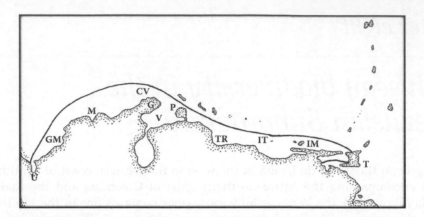

Figure 8.1 Map of the Venezuelan Subprovince of the Caribbean Province, showing some of the important geographical features. U = Golfo de Urabá; GM = Golfo de Morrosquillo; M = Mouth of the Magdalena River; CV = Cabo La Vela; G = Goajira Peninsula; V = Golfo de Venezuela; P = Paraguaná Peninsula; TR = Golfo Triste; IT = Islas Testigos; IM = Isla Margarita; T = Trinidad.

Using the raw data listed in Appendix 1 and based on the percentages of endemism in the 10 key index taxa found on these four environmental regions, the Venezuelan Subprovince can be seen to be a discrete biogeographical entity. Applying the relationship

$$t = \frac{n}{N}(100), \ t > 25$$

these are as follows:

Modulidae ($t1$), $t = 0$, with $N = 1$, $n = 0$
Turbinellidae ($t2$), $t = 0$, with $N = 3$, $n = 0$
Conidae ($t3$), $t = 69$, with $N = 29$, $n = 20$
Conilithidae ($t4$), $t = 0$, with $N = 4$, $n = 0$
Muricinae ($t5$), $t = 57$, with $N = 19$, $n = 11$
Fasciolariinae ($t6$), $t = 66$, with $N = 3$, $n = 2$
Lyriinae ($t7$), $t = 0$, with $N = 1$, $n = 0$
Olivinae ($t8$), $t = 50$, with $N = 6$, $n = 3$
Cancellariinae ($t9$), $t = 33$, with $N = 3$, $n = 1$
Plesiotritoninae ($t10$), $t = 50$, with $N = 2$, $n = 1$

The subprovincial combined index, incorporating the endemicity of these 10 families and subfamilies and applying the relationship

$$S = \sum_{n=1}^{10} \frac{tn}{10}, \ S > 25$$

yields the quantity $S = 34.1$. This high level of subprovincial endemism ($S > 25$) demonstrates that the Venezuelan Subprovince is a strong and well-demarcated subdivision of the Caribbean Province.

Figure 8.2 View of the deck and crew of the Vikingos de Colombia S.A. commercial shrimp boat *Cartagenera*, showing the sea life that was collected after a three-hour dredge tow in the Golfo de Morrosquillo, Colombia. The jumbled mass of benthonic organisms included numerous penaeid shrimp, along with crabs, echinoderms, sponges, fishes, and red algae clumps. The gastropods *Lindaconus phlogopus*, *Voluta virescens*, *Vokesimurex bayeri*, *Sconsia lindae*, and *Oliva (Americoliva) bewleyi* were commonly encountered in these trawls in the Golfo de Morrosquillo area off Tolu. The crew members were also very pleased by the retrieval of a drowned green sea turtle, the source of fresh meat for the next few meals. Sadly, many sea turtles are caught in shrimp nets and drown in this fashion. (From E. Petuch, archival photograph, 1976.)

Golfo de Morrosquillo and Colombian coast

The area extending from the Golfo de Urabá on the Panamá–Colombia Border to the mouth of the Magdalena River near Barranquilla and encompassing the Golfo de Morrosquillo and the coastal lagoons near Cartagena is characterized as having muddy, organic-rich sea floors. These soft substrate–reducing environments receive an overabundance of nutrients from the effluent of the Atrato, Sinu, and Magdalena rivers and the immense marshlands of the Bahia de Barbacoas near Cartagena. A distinctive and characteristic malacofauna has evolved on these muddy substrates, with one of the most noteworthy being a large fauna of nuculanid bivalves. This group of thin-shelled, elongated bivalves (Family Nuculanidae) normally occurs in offshore, deep-water areas around the Caribbean and is represented by a few rare species. In the area of Colombia between the Golfo de Morrosquillo and the Magdalena River mouth, however, nuculanids are found abundantly at intertidal and shallow subtidal depths (Petuch, 1987: 116, 117, 138), with a least four species having been collected: the large, hatchet-shaped *Adrana notabilis*; the elongated *Adrana scaphoides*; the

Figure 8.3 Endemic mollusks of the northern Colombian coast: (A) *Lindaconus phlogopus* (Tomlin, 1937), length 65 mm. (B) *Adrana (Scissuladrana) ludmillae* (Petuch, 1987), length 40 mm. (C) *Vokesimurex bayeri* Petuch, 2001, length 100 mm. (D) *Oliva (Americoliva) bewleyi* Marrat, 1870, length 59 mm. (E) *Gradiconus cingulatus* (Lamarck, 1810), holotype, length 44 mm. (F) *Tenorioconus granarius* (Kiener, 1845), length 45 mm.

small *Adrana tellinoides*; and the cross-hatched sculptured *Adrana (Scissuladrana) ludmillae* (Figure 8.3B). This unusual, regionally restricted nuculanid radiation occurs along with a highly endemic gastropod fauna, including the following (several illustrated in Figures 8.3 and 8.4):

Cassidae
    Sconsia lindae (found in depths as shallow as 3 m; a living member of the Pliocene
        shallow-water *Sconsia laevigata* species complex)
Ficidae
    Ficus lindae
Muricidae
    Panamurex eugeniae
    Vokesimurex bayeri
    Vokesimurex olssoni (also found along Costa Rica and Panamá)
Volutidae
    Scaphella evelina (deeper water areas)
Marginellidae
    Prunum poulosi

Olividae
   Oliva (Americoliva) bewleyi
   Oliva (Americoliva) obesina
   Oliva (Americoliva) oblonga
Conidae
   Dauciconus vikingorum
   Lindaconus phlogopus
   Poremskiconus colombianus (Rosarios Islands near Cartagena)
   Purpuriconus alainalaryi (Rosarios Islands near Cartagena)
   Tenorioconus granarius

Along with these endemic species, the offshore muddy environments also house a large contingent of typical southern Caribbean gastropods such as *Strombus pugilis* (Strombidae), *Chicoreus brevifrons* (Muricidae), *Melongena melongena* (Melongenidae), and *Voluta virescens* (Volutidae). The shallow muddy intertidal areas within the Bahia de Barbacoas and the muddy coastal lagoons near Cartagena also harbor a number of interesting endemic gastropods, including the following:

Muricidae
   Vokesimurex cf. messorius (inflated body whorl with brown band and short siphonal canal)
   Vokesimurex cf. woodringi (found near mangroves in the Cartagena Lagoons)
Volutidae
   Voluta virescens (classic dwarf shallow water form)
Marginellidae
   Prunum marginatum
Conidae
   Gradiconus cingulatus (abundant in the intertidal areas at La Boquilla, Cartagena)
Conilithidae
   Perplexiconus puncticulatus form *mauritianus*
Terebridae
   Strioterebrum peteveriana

Endemism along the Goajira Peninsula

The Goajira Peninsula, the northernmost piece of land in Colombia, is a burning desert that is exposed to a near constant strong wind. Blowing from east to west, the powerful trade winds scour the Goajira area and prevent large-scale precipitation from developing. The entire coastline of the peninsula is inhospitable, composed of low, jagged rocks and dense forests of cacti and thorny acacia scrub. The strong, continuously blowing wind pushes the surface water away from the shore and, through entrainment, pulls deep, nutrient-rich water up to the surface. These upwellings fertilize the phytoplankton in the euphotic zone and create almost continuous plankton blooms and high-productivity surface water conditions. These upwelling and high-productivity areas are especially well developed near Cabo La Vela, on the northwestern side of the Goajira Peninsula. Consequently, immense quantities of penaeid shrimp swarm to this area, and Cabo La Vela is one of the principal destinations for the commercial shrimp fleets.

   Unlike the areas south of the Magdalena River, the sea floor off the Cabo La Vela area is composed of coarse shell hash and is covered with dense growths of large sponges. While working on the shrimp boats off Cabo La Vela, I had many personal encounters

Figure 8.4 Endemic gastropods of the northern Colombian coast: (A) *Ficus lindae* Petuch, 1988, length 85 mm. (B) *Oliva (Americoliva) bayeri* Petuch, 2001, holotype, length 36.5 mm. (C) *Dauciconus goajira* (Petuch, 1992), holotype, length 35 mm. (D) *Oliva (Americoliva) goajira* Petuch and Sargent, 1986, length 43 mm. (E) *Dauciconus vikingorum* (Petuch, 1992), holotype, length 37 mm. (F) *Tenorioconus jesusramirezi* (Cossignani, 2010), length 34 mm. (G) *Vokesimurex olssoni* (E. Vokes, 1967), length 50 mm (also found along Panama).

with the most dreaded of all sponges, the *Neofibularia nolitangere* (the Touch-Me-Not Sponge). These formed entire bioherms, which were hauled up onto the deck along with the shrimp and other captured sea life. Upon contact with skin, these relatively insipid-appearing sponges produced an instant intense stinging and burning, which often lasted for hours. Unfortunately, the large and beautiful endemic cowrie *Muracypraea donmoorei* (Figure 8.5C,D) lives among these stinging sponges and feeds on their poison-laden tissues with impunity. Farther to the north along the peninsula, and in deeper water of over 100 m, another endemic cowrie lives on these sponge bioherms, the large rat-eared *Muracypraea tristensis* (Figure 8.5E,F). Both *Muracypraea tristensis* and the shallower water *Muracypraea donmoorei* are restricted to the small geographical area of the Goajira Peninsula, Golfo de Venezuela, and Golfo Triste and represent relicts from the Pliocene Epoch. At depths of 35m to 100 m off the Goajira Peninsula, the *Neofibularia* and other species of sponges form the base for an entire endemic ecosystem, filled with molluscan species that are found nowhere else (see Petuch, 1987, for discussions and illustrations of the Goajira mollusks). Some of the more prominent and important Goajira Peninsula gastropods include (several illustrated in Figures 8.4 and 8.5):

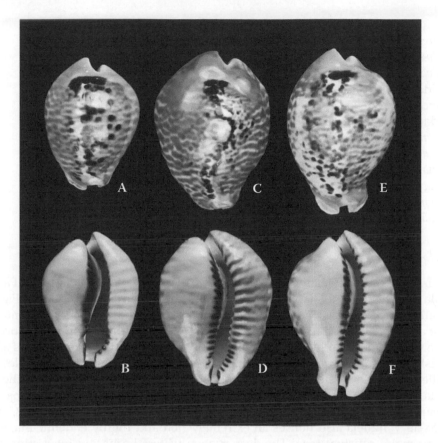

Figure 8.5 Relictual cowries from the Venezuelan Subprovince: (A, B) *Muracypraea mus* (Linnaeus, 1758), length 42 mm. (C, D) *Muracypraea donmoorei* (Petuch, 1979), length 58 mm. (E, F) *Muracypraea tristensis* (Petuch, 1987), length 61 mm.

Cypraeidae
  Muracypraea donmoorei (also found in the adjacent Golfo de Venezuela)
  Muracypraea tristensis (also found in the adjacent Golfo de Venezuela)
Ovulidae
  Turbovula lindae
Muricidae
  Panamurex harasewychi
  Vokesimurex sunderlandi
Buccinidae
  Antillophos bayeri
Colubrariidae
  Colubraria bayeri
Costellariidae
  Turricostellaria lindae
Volutomitridae
  Volutomitra erebus

Olividae
  Eburna glabrata (also found in the Golfo de Venezuela)
  Oliva (Americoliva) bayeri
  Oliva (Americoliva) goajira
Harpidae-Moruminae
  Cancellomorum lindae
Conidae
  Attenuiconus poulosi
  Conasprelloides penchaszadehi
  Conasprelloides velaensis
  Dauciconus goajira
  Tenorioconus jesusramirezi
  Tenorioconus mappa sanguineus

Golfo de Venezuela

Almost landlocked between the Goajira Peninsula of Colombia and the Paraguaná Peninsula of Venezuela, the Golfo de Venezuela forms a veritable inland sea (Figure 8.6). This small body of water is oceanographically complex, with fresh water effluent entering from the attached Lake Maracaibo in the south; with large hypersaline evaporite pools along the Golfete de Coro in the east; with cool water, nutrient-rich upwellings along the Monjes Islands and the mouth of the Gulf in the north; and with large, shallow turtle-grass–filled bays along the western side of the Paraguaná Peninsula. Like the Goajira Peninsula, the area surrounding the Golfo de Venezuela is a desert and is subject to the same nearly perpetual strong trade winds. As a microcosm of the southern Caribbean, the Golfo de Venzuela also houses an extremely rich and highly endemic malacofauna, containing many relictual taxa. The deeper open waters harbor an interesting endemic species radiation of the conid genus *Conasprelloides*, containing at least four species. Some of the more important of the gastropods from the deeper central area of this enclosed sea include the following (with several illustrated in Figures 8.3, 8.5, 8.7, 8.9, 8.10, and 8.11):

Cypraeidae
  Muracypraea donmoorei
  Muracypraea tristensis
Ovulidae
  Pseudocyphoma gibsonsmithorum
Muricidae
  Panamurex velero (also found in the Golfo Triste)
  Vokesimurex cf. messorius
Fasciolariidae
  Fusinus ansatus (= *closter caboblanquensis*)
Columbellidae
  Strombina cf. caboblanquensis
  Strombina pumilio
Costellariidae
  Turricostellaria leonardhilli
Olividae
  Amalda tankervillei (also found in the Golfo Triste and Isla Margarita)
  Turrancilla williamsoni

Figure 8.6 View of the Golfo de Venezuela from near the village of Amuay, Falcón State, Venezuela, showing the arid desert conditions of the Paraguaná Peninsula. The large Cardones cactus plants shown here often form impenetrable forests along the shoreline areas. Commercial shrimpers plying the waters off this desert coast have trawled up many rarely seen endemic gastropods, including *Muracypraea tristensis, Turrancilla williamsoni, Gradiconus parascalaris*, and *Attenuiconus honkeri*. (From E. Petuch, archival photograph, 1979.)

Conidae
   Attenuiconus honkeri
   Conasprelloides finkli
   Conasprelloides kevani
   Conasprelloides tristensis (also found in the Golfo Triste)
   Conasprelloides venezuelanus
   Gradiconus gibsonsmithorum
   Gradiconus parascalaris
   Gradiconus paulae
Turridae
   Polystira lindae

Probably the most interesting area within the Golfo de Venezuela is Amuay Bay and environs, near Judibana, on the western side of the Paraguaná Peninsula. Here, the wide, shallow bay is filled with immense beds of turtle grass (*Thalassia testudinum*) that cover at least half of the sea floor. Interspersed among the sea grass beds are large patches of open bottom that are literally paved by dense beds of the large (av. 110 mm) turritellid gastropod

Figure 8.7 Venezuelan Subprovince shallow-water volutes: (A) *Voluta musica* Linnaeus, 1758, length 44 mm. (B) *Voluta musica guinaica* Lamarck, 1811, length 60 mm. (C) *Voluta virescens* Lightfoot, 1786, length 43 mm. (D) *Voluta viresecens* form *fulva* Lamarck, 1911, length 83 mm.

Broderiptella variegata paraguanensis. The turtle grass beds and turritellid beds form the base of a relictual molluscan assemblage that represents an almost intact late Pliocene–Pleistocene community (Petuch, 1976; Vermeij, 1978). Of special interest in the turtle grass beds is the mouse cowrie, *Muracypraea mus*, which feeds on the epibiota growing on the grass blades and on the sea grass blades themselves (Figure 8.8). In direct contrast, the other closely related deeper water *Muracypraea* species, *M. donmoorei* and *M. tristensis*, are carnivorous and feed on sponges (Petuch, 1979, 1988). Often occurring in immense aggregations of thousands of individuals, the mouse cowrie prefers the shallowest intertidal areas and is semiamphibious, crawling up at night onto the exposed mud flats. *Muracypraea mus* ranges from Riohacha, on the Goajira Peninsula near Cabo La Vela, all the way to Los Taques on the northern tip of the Paraguaná Peninsula but is most abundant in Amuay Bay. The suspension-feeding turritellids are the main prey item for several large molluscivorous gastropods, including the muricids *Phyllonotus* cf. *margaritensis* and *Chicoreus brevifrons*, the fasciolariid *Fasciolaria tulipa hollisteri*, and the volutid *Voluta musica guinaica*. Some of the more prominent molluscan members of the turtle-grass–turritellid community in Amuay Bay include the following (several illustrated in Figures 8.5, 8.7, and 8.9):

Turritellidae
    Broderiptella variegata paraguanensis
Cypraeidae
    Muracypraea mus

Figure 8.8 Close-up view of two living mouse cowries, *Muracypraea mus* (Linnaeus, 1758), crawling in mud and *Enteromorpha* algae in Amuay Bay, Paraguaná Peninsula, Venezuela. Both specimens are approximately 40 mm in length. Note the white siphon, black tentacles, and warty mantle (mostly retracted) of the living animal. (Photograph by M. G. Harasewych, who took the picture while accompanying the author on a research trip to Amuay Bay in 1979.)

Muricidae
  Chicoreus brevifrons
  Phyllonotus cf. margaritensis (probably a new species)
  Vokesimurex cf. messorius
Fasciolariidae
  Fasciolaria tulipa hollisteri
Volutidae
  Voluta musica guinaica
Olividae
  Eburna glabrata
  Oliva (Americoliva) fulgurator
  Oliva (Americoliva) porcea (also found along the adjacent Goajira Peninsula)
  Oliva (Americoliva) reclusa subspecies
Conidae
  Gradiconus undatus
  Lindaconus baylei (= arubaensis)
Conilithidae
  Perplexiconus puncticulatus (form *cardonensis*)

The combined area of the Goajira and Paraguaná Peninsulas and Golfo de Venezuela is now known to be geographically heterochronous and is a classic example of a secondary

Figure 8.9 Relictual taxa of the Venezuelan Subprovince: (A) *Phyllonotus cf. margaritensis* (Abbott, 1958), length 60 mm (common in the Pleistocene fossil beds of Venezuela). (B) *Strombina pumilio* (Reeve, 1859), length 16 mm (common in the Pleistocene fossil beds of Venezuela). (C) *Cancellomorum lindae* (Petuch, 1987), length 35 mm (a living descendant of the Miocene and Pliocene *Cancellomorum dominguense* species complex). (D) *Sconsia lindae* Petuch, 1988, length 70 mm (a living shallow water descendant of the Miocene and Pliocene *Sconsia laevigata* species complex). (E) *Eburna glabrata* (Linnaeus, 1758), length 75 mm (common in the Pleistocene fossil beds of Venezuela). (F) *Mazatlania cosentini* (Philippi, 1836), length 22 mm (common in the Pleistocene fossil beds of Venezuela). (G) *Oliva (Americoliva) porcea* Marrat, 1870, length 57 mm (a living descendant of the Pleistocene Venezuelan fossil *Oliva (Americoliva) schepmani*).

relict pocket (Petuch, 1982, 1988). Many of the genera and actual species of gastropods found within this area are components of molluscan assemblages that have changed very little since the late Pliocene and early Pleistocene (see Weisbord, 1962, for illustrations of the Pliocene–Pleistocene fossil gastropods of Venezuela). Some of the Golfo de Venezuela living species that are also known from the Pleistocene fossil beds of Venezuela include the columbellids *Strombina pumilio* and *Mazatlania cosentini* (Figure 8.9), the muricid *Phyllonotus cf. margaritensis* (referred to as *Phyllonotus globosus*, which is actually a different Pliocene species from the eastern United States), the cypraeid *Muracypraea donmoorei*, the turritellid *Broderiptella variegata paraguanensis*, and the volutid *Voluta musica guinaica* (all illustrated as fossils in Weisbord, 1962).

Although having a similar molluscan fauna to those of the Golfo de Venezuela and Golfo Triste, Margarita Island and the neighboring Isla Coche, Isla Cubagua, and Los Testigos Islands also have a few interesting shallow water endemic gastropods. Primary among these is the cone shells *Tenorioconus trinitarius* and *Tenorioconus caracanus*, the neritid

Figure 8.10 Endemic cone shells of the Venezuelan Subprovince: (A) *Attenuiconus poulosi* (Petuch, 1992), holotype, length 36 mm. (B) *Tenorioconus mappa sanguineus* (Kiener, 1845), holotype, length 43 mm. (C) *Gradiconus paulae* (Petuch, 1988), holotype, length 27 mm. (D) *Attenuiconus honkeri* (Petuch, 1988), holotype, length 37 mm. (E) *Conasprelloides velaensis* (Petuch, 1992), holotype, length 31 mm. (F) *Conasprelloides penchaszadehi* (Petuch, 1986), holotype, length 18 mm (the holotype is a juvenile specimen, as adult shells reach lengths of 35 mm).

Nerita amplisulcata, and the large muricid *Phyllonotus margaritensis* (the true *margaritensis*, which differs from the *globosus* type of the Golfo de Venezuela). The cone shells of these areas are shown in Figures 8.10 and 8.11.

Venezuelan deep-water areas

To the east of the Paraguaná Peninsula, the deeper water areas in the Golfo Triste and off the adjacent coastline have also proven to be of special interest. These deep-water (100–300 m depths) sea floors, covered with fine terriginous sediments, have been shown to harbor a highly endemic malacofauna with a large number of relictual taxa. As seen along the Golfo de Venezuela and Goajira Peninsula, the Golfo Triste also is under the influence of cool, nutrient-rich upwelling systems and supports a large commercial shrimp fishing industry. Collections of mollusks taken by these deep-water shrimpers are the sole source of information about the composition of the offshore malacofaunas. Of particular interest in these deep-water areas was the discovery of a living member of the odd clath-urellid genus *Paraborsonia*, a group previously thought to have become extinct at the end of the Miocene. This relictual taxon was also found to occur along with a member of the buccinid genus *Truncaria*, a group previously known from only the eastern Pacific Panamic

Figure 8.11 Endemic cone shells of the Venezuelan Subprovince: (A) *Conasprelloides venezuelanus* (Petuch, 1987), holotype, length 27 mm. (B) *Conasprelloides kevani* (Petuch, 1987), holotype, length 17 mm. (C) *Conasprelloides finkli* (Petuch, 1987), length 28 mm. (D) *Conasprelloides tristensis* (Petuch, 1987), length 45 mm. (Photo by A. Poremski.) (E) *Gradiconus parascalaris* (Petuch, 1987), holotype, length 23 mm. (F) *Gradiconus gibsonsmithorum* (Petuch, 1986), holotype, length 18 mm. (G) *Tenorioconus caracanus* (Hwass, 1792), length 38.6 mm, off Margarita Island.

Province (Petuch, 1987). Other unusual or endemic deep-water taxa found of the Golfo Triste and adjacent areas include the following (several illustrated in Figure 8.12):

Turbinidae
  Categeis finkli
Muricidae
  Panamurex petuchi
Fasciolariidae
  Harasewychia harasewychi
Columbariidae
  Fulgurofusus brayi
Buccinidae
  Truncaria lindae
Mitridae
  Subcancilla leonardhilli
  Subcancilla lindae

Figure 8.12 Endemic deep-water species of the Venezuelan Subprovince: (A) *Harasewychia harasewychi* Petuch, 1987, holotype, length 34 mm. (B) *Cateigis finkli* (Petuch, 1987), holotype, length 18 mm. (C) *Paraborsonia lindae* Petuch, 1987, length 16 mm. (D) *Vokesimurex sunderlandi* (Petuch, 1987), holotype, length 41 mm. (E) *Turbovula lindae* (Petuch, 1987), holotype, length 17 mm. (F) *Nodicostellaria kremerae* Petuch, 1987, holotype, length 18 mm. (G) *Pseudocyphoma gibsonsmithorum* Petuch, 1987, holotype, length 17 mm.

Costellariidae
    Nodicostellaria kremerae
Conidae
    Conasprelloides tristensis (also found in the Golfo de Venezuela)
Clathurellidae
    Paraborsonia lindae

chapter nine

Molluscan biodiversity in the Grenadian and Surinamian Subprovinces

Extending from Aruba in the west to Anguilla in the north, and encompassing the coral islands off Venezuela, northern Trinidad, and the Leeward and Windward Islands of the Lesser Antilles, the Grenadian Subprovince (Figure 9.1) is composed entirely of island archipelagos. Within the subprovincial boundaries, five separate centers of speciation can be differentiated. These include: (1) the Dutch Islands of Aruba, Curaçao, and Bonaire (the ABC Islands); (2) the coralline islands and atolls off Venezuela; (3) the Lesser Antilles Islands from Tobago to Guadeloupe; (4) the Lesser Antilles Islands from Guadeloupe to Anguilla; and (5) the Barbados Seamount. Since most of the island chains, particularly those in the Windward Group, are separated by very deep water channels and trenches, many nonvagile mollusks have become genetically isolated and have undergone speciation or incipient speciation (subspeciation). Several groups of gastropods have evolved endemic species radiations along some of the island chains, with the conid genus *Tenorioconus* being one of the most prominent and characteristic having evolved at least seven species and subspecies in the area extending from Guadeloupe to Aruba (discussed later in this chapter). The Grenadian Subprovince also contains several endemic genera, including *Roquesia* (Muricidae; see Appendix 2), *Globivasum* (Turbinellidae), *Aphera* (Cancellariidae), and *Arubaconus* (Conidae; see Appendix 2), and endemic species radiations of the genera *Oliva (Americoliva)* (Olividae) and *Dauciconus* (Conidae).

Utilizing the raw data listed in Appendix 1 and based on the percentages of endemism in the 10 key index taxa found in these five centers of speciation, the Grenadian Subprovince can be seen to be a discrete biogeographical entity within the Caribbean Province. Applying the relationship

$$t = \frac{n}{N}(100), \, t > 25$$

these are the following:

Modulidae (*t*1), *t* = 0, with *N* = 2, *n* = 0
Turbinellidae (*t*2), *t* = 40, *N* = 5, *n* = 2
Conidae (*t*3), *t* = 70, *N* = 27, *n* = 19
Conilithidae (*t*4), *t* = 36, *N* = 11, *n* = 4
Muricinae (*t*5), *t* = 28, with *N* = 21, *n* = 6
Fasciolariinae (*t*6), *t* = 0, with *N* = 1, *n* = 0
Lyriinae (*t*7), *t* = 100, with *N* = 4, *n* = 4
Olivinae (*t*8), *t* = 55, *N* = 11, *n* = 6
Cancellariinae (*t*9), *t* = 25, with *N* = 4, *n* = 1
Plesiotritoninae (*t*10), *t* = 0, with *N* = 1, *n* = 0

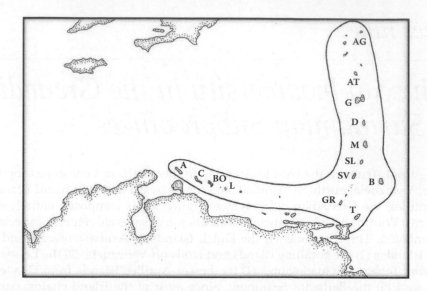

Figure 9.1 Map of the Grenadian Subprovince of the Caribbean Province, showing some of the important geographical features: A = Aruba; C = Curaçao; BO = Bonaire; L = Los Roques Atoll; GR = Grenada; T = Trinidad and Tobago; B = Barbados; SV = Saint Vincent; SL = Saint Lucia; M = Martinique; D = Dominica; G = Guadeloupe; AT = Antigua and Barbuda; AG = Anguilla.

The Subprovincial Combined Index, incorporating the endemicity of these 10 families and subfamilies and based on the relationship

$$S = \sum_{n=1}^{10} \frac{tn}{10}, \; S > 25$$

yields the quantity S = 48.3. This very high level of subprovincial endemism (S>25) demonstrates that the Grenadian Subprovince is a strong and well-defined subdivision of the Caribbean Province.

Lesser Antilles and Grenadines

In a narrow curving arc running in a north–south direction, the Lesser Antilles extend from Grenada northward to Anguilla, and include large islands such as Saint Vincent, Saint Lucia, Martinique, Dominica, Guadeloupe, and Antigua and hundreds of smaller rocky and coralline cays and islets (Figure 9.2). The wide and deep Anegada Passage between the Virgin Islands and Anguilla, with its strong currents, acts as a faunal barrier between the Antillean and Grenadian Subprovinces. Likewise, the wide Windward Passage and deep Grenada Basin between Grenada and Trinidad and the mainland of Venezuela act as faunal barriers between the Venezuelan and Grenadian Subprovinces. The two wide, deep passages segregate the Lesser Antilles from the rest of the Caribbean and have kept the islands isolated since their inception during the Pliocene Epoch. Because of over 3 million years of genetic isolation of the resident nonvagile taxa, the Lesser Antilles exhibit a high degree of endemism in their molluscan faunas. Principal among these, as mentioned previously, is the cone shell genus *Tenorioconus*, with six species (or possible

Figure 9.2 View of Saline Island, Grenadian Grenadines, from the southern tip of Carriacou Island, showing the extensive system of coral reefs. Areas like this are typical of both the Vincentian and Grenadian Grenadines. On the reefs just offshore, several classic endemic Grenadian Subprovince shells were commonly collected, including the conids *Poremskiconus beddomei* and *Tenorioconus pseudoaurantius*. (From E. Petuch, archival photograph, 1979.)

subspecies) that are confined to the southern part of the archipelago, from Dominica in the north to Grenada in the south. These include the following (several illustrated on Figure 9.3):

Family Conidae
   Tenorioconus cedonulli (endemic to Saint Vincent)
   Tenorioconus dominicanus (endemic to Dominica)
   Tenorioconus grenadensis (endemic to Grenada and the Grenadines)
   Tenorioconus insularis (endemic to Saint Lucia)
   Tenorioconus martinicanus (endemic to Martinique)
   Tenorioconus pseudoaurantius (endemic to Grenada and the Grenadines)

These spectacularly colored cone shells have been much sought after by collectors for over 300 years and they are considered to be some of the most desirable shells in the entire Caribbean area. All six species live in relatively shallow water, in depths ranging from 1–50 m, and prefer sandy areas adjacent to turtle grass beds and patches of coral rubble. As in many of the Caribbean cones, the genus *Tenorioconus* has direct development larvae that lack a planktonic stage and the adults remain close to where they were born for their entire lives. Being shallow-water, nondispersing gastropods, each species inhabits a narrow ring of habitats that surrounds its resident island. Since the islands are all separated by very deep water, the shallow water, nonvagile cones shells are essentially trapped on each island and have undergone genetic drift and speciation. This same evolutionary pattern is seen in the terrestrial environments of the Lesser Antilles, where each island has

Figure 9.3 Cone shells of the Lesser Antilles: (A) *Poremskiconus hennequini* (Petuch, 1992), holotype, length 17.5 mm. (B) *Tenorioconus dominicanus* (Hwass, 1792), length 41 mm. (C) *Tenorioconus pseudo-aurantius* Vink and von Cosel, 1985, holotype, length 34.5 mm. (D) *Tenorioconus grenadensis* (Hwass, 1792), length 40 mm. (E) *Dauciconus boui* (da Motta, 1988), holotype, length 30.2 mm. (F) *Tenorioconus cedonulli* (Linnaeus, 1767), length 43 mm. (Photo by A. Poremski.)

evolved its own resident *Anolis* iguanid lizard and Amazon parrot, both from groups that are not easily dispersable.

Besides the *Tenorioconus* species complex, many other groups of gastropods have undergone similar genetic isolation and have evolved species complexes. Of particular interest is the Grenadian Subprovince endemic genus *Globivasum* (Turbinellidae), with *Globivasum globulus* being confined to the shallow Antigua and Barbuda Islands bank and with a subspecies, *Globivasum globulus whicheri*, occurring only on the isolated bank around Anguilla Island. The two shallow banks and island groups are separated by a deep channel, which acts as a barrier to dispersal. The Island of Martinique has also proven to be a center of speciation for the Conidae, with two endemic *Dauciconus* species (*D. boui* and *D. norai*; Figures 9.3E and 9.4A) and two endemic *Poremskiconus* species (*P. hennequini* and *P. colombi*; Figures 9.3A and 9.4G) occurring on coral reefs. The coral reef areas of the central Lesser Antilles islands of Guadeloupe, Dominica, and Martinique are also the habitat of the spectacular endemic muricid, *Pterynotus phyllopterus* (Figure 9.4E), a relictual species with Miocene affinities. Grenada and the Grenadines Islands, particularly Carriacou Island, also harbor several endemic species, including the cone shells *Poremskiconus beddomei* and *Tenorioconus pseudoaurantius*. Some of the more important Lesser Antillean endemic gastropod taxa include (several illustrated here on Figures 9.3 and 9.4):

Figure 9.4 Endemic gastropods of the Grenadian Subprovince: (A) *Dauciconus norai* (da Motta and Raybaudi, 1992), holotype, length 37 mm. (B) *Oliva (Americoliva) fulgurator fusiforme* Lamarck, 1811, length 49 mm (endemic to Aruba). (C) *Jaspidiconus jaspideus* (Gmelin, 1791), length 22 mm (restricted to the Grenadian and Venezuelan Subprovinces; shown here for comparison with other Carolinian, Caribbean, and Brazilian Province *Jaspidiconus* species). (D) *Enaeta guildingi* (Sowerby, 1844), length 14 mm. (E) *Pterynotus phyllopterus* (Lamarck, 1822), length 73 mm. (F) *Jaspidella carminiae* Petuch, 1992, length 25 mm. (G) *Poremskiconus colombi* (Monnier and Limpalaer, 2012), length 18 mm (endemic to Martinique).

Cassidae
   Sconsia nephele
Muricidae
   Hexaplex strausi (endemic to Martinique and Dominica; the only *Hexaplex* in the Caribbean)
   Pterynotus phyllopterus (endemic to Guadeloupe, Dominica, and Martinique)
Turbinellidae
   Globivasum globulus (endemic to Antigua and Barbuda)
   Globivasum globulus whicheri (endemic to Anguilla; see Appendix 2)
Volutidae
   Enaeta guildingi
   Lyria archeri (endemic to Montserrat, Guadeloupe, and Martinique)
   Lyria beaui
   Lyria sabaensis (endemic to Saba)

Olivellidae
   Jaspidella carminiae (also found on the Los Roques Atoll and Tobago)
   Oliva (Americoliva) olivacea (endemic to Grenada and the Grenadines; see Appendix 2)
   Oliva (Americoliva) drangae (endemic to Trinidad and Tobago)
Conidae
   Dauciconus boui (endemic to Martinique)
   Dauciconus norai (endemic to Martinique)
   Poremskiconus beddomei (endemic to Saint Vincent, Grenada and the Grenadines)
   Poremskiconus colombi (endemic to Martinique)
   Poremskiconus hennequini (endemic to Martinique)

Endemism on the Dutch ABC Islands and Los Roques Atoll

Encompassing the Islands of Aruba, Curaçao, Bonaire, Klein Bonaire, and several small islets, the Dutch ABC Islands (now two separate countries) occupy the farthest western edge of the Grenadian Subprovince. The Island of Aruba lies only 40 km off the Paraguaná Peninsula of mainland Venezuela and actually occupies the outer edge of the Venezuelan continental shelf. Because of this, Aruba is under the influence of the upwelling systems that occur off the mouth of the Golfo de Venezuela and has continental-type marine conditions, including high-productivity, plankton-rich water. Curaçao and Bonaire, on the other hand, are farther offshore and are surrounded by deep water, with much more sterile, open-oceanic marine conditions. Of the three main islands, Aruba has the highest level of endemism, with several gastropods that are found nowhere else in the Caribbean. Some of these Aruban endemics include (several illustrated on Figure 9.5):

Muricidae
   Murexiella hilli
Olividae
   Eburna balteata
   Oliva (Americoliva) fulgurator fusiforme
   Oliva (Americoliva) reclusa
   Oliva (Americoliva) sargenti
Conidae
   Arubaconus hieroglyphus (see Appendix 2)
   Tenorioconus curassaviensis

Several of these Aruban endemics, particularly *Arubaconus hieroglyphus*, *Oliva (Americoliva) sargenti*, and *Murexiella hilli*, are very limited in their geographical ranges and appear to be restricted to a small stretch of coastline near Malmok. Here, the yellow mussel, *Brachidontes modiolus*, forms extensive beds in shallow water and these are the sole habitat of the mussel-eating *Murexiella hilli*. In the turtle grass beds of the Malmok coast, several typical Golfo de Venezuela gastropods are also present and these occur together with the Aruban endemics, forming a distinctive and unusual molluscan assemblage. Some of these Venezuelan coastal taxa include the cone shell *Lindaconus baylei* (= *L. spurius arubaensis*) and the olivids *Eburna glabrata* and *Oliva (Americoliva) reclusa*. The islands of Curaçao and Bonaire exhibit far less endemism than does Aruba, with only one well-known restricted taxon, the cone shell *Tenorioconus aurantius* (Figure 9.5F). Bonaire and Curaçao also lack the Venezuelan coastal species that are found on Aruba and each has a classic Lesser Antillean shallow water reef-associated malacofauna.

Figure 9.5 Endemic gastropods from the ABC Islands and Los Roques Atoll: (A) *Oliva (Americoliva) sargenti* Petuch, 1988, length 32 mm. (endemic to Aruba). (B) *Tenorioconus curassaviensis* (Hwass, 1792), length 45 mm. (Photo by A. Poremski.) (C) *Arubaconus hieroglyphus* (Duclos, 1833), length 12 mm. (endemic to Aruba). (D) *Murexiella hilli* Petuch, 1987, length 37 mm. (endemic to Aruba). (E) *Tenorioconus duffyi* (Petuch, 1992), holotype, length 39 mm. (endemic to Los Roques). (F) *Tenorioconus aurantius* (Hwass, 1792), length 49 mm. (Photo by A. Poremski.)

Immediately to the east of the ABC Islands are three large, low coralline island groups; Las Aves Islands, Los Roques Atoll, and La Orchila Island. Of these three, Los Roques Atoll is by far the largest and most complex oceanographically. One of only 13 atolls in the western Atlantic, Los Roques contains 350 coral islands and covers over 221,120 hectares. Within this isolated labyrinth of coral cays, several endemic gastropods have evolved, including the cone shell *Tenorioconus duffyi* (Figure 9.5E) (Petuch, 1992c) and the ergalitaxine muricid *Roquesia lindae* (new genus and new species). This tiny new murex shell (Figure 9.6A,B) more closely resembles members of the northern Caribbean genus *Minibraria* than it does any other known Atlantic ergalitaxine group (see Appendix 2). The large, slender olive shell, *Oliva (Americoliva) fulgurator bullata*, a Los Roques endemic relative of the Aruban endemic *Oliva (Americoliva) fulgurator fusiforme*, is abundant on open coral sand bottoms within the Los Roques lagoon, as is the Grenadian Subprovince endemic olivellid *Jaspidella carminiae* and the endemic pure white dwarf donacid bivalve

Figure 9.6 New discoveries from the Grenadian Subprovince: (A, B) *Roquesia lindae* Petuch, new genus and new species, holotype, length 9 mm. (C, D) *Globivasum globulus whicheri* Petuch, new subspecies, holotype, length 29 mm. (E) *Globivasum globulus* (Lamarck, 1816), length 26 mm, Antigua Island. (for comparison with *G. globulus whicheri*). (F, G) *Dalliconus colletteae* new species, holotype, length 21 mm. The descriptions of these new taxa are given in Appendix 2.

Donax denticulatus form *stephaniae* (see Petuch, 1992). Cone shells similar to *Tenorioconus duffyi* have been collected on the neighboring Las Aves Islands, and these may represent either a new, unnamed species or a localized variety of the Los Roques *T. duffyi*. Some of the mollusks that are endemic to, or are especially common on, the Los Roques Atoll and adjacent coral islands include the following (two shown in Figures 9.4 and 9.5):

Gastropoda
Muricidae
  Muricopsis huberti (= *duffyi*; also found throughout the Lesser Antilles)
  Roquesia lindae (Appendix 2)
Columbellidae
  Zaphrona lindae (also found in northern Brazil)
Olividae
  Oliva (Americoliva) fulgurator bullata
Conidae
  Tenorioconus duffyi

Bivalvia
Donacidae
   Donax denticulatus form *stephaniae*

Endemism on Barbados

The Barbados Seamount, which rises abruptly out of the depths of the Tobago Basin, is geologically unrelated to the Lesser Antilles Island Chain. Unlike the Lesser Antilles, which are geologically young volcanic peaks dating from the Miocene and Pliocene Epochs, Barbados is an old island, dating from at least the Cretaceous. During the early Cenozoic, the Barbados Seamount was dragged through the then-open Panamá area from the Pacific Ocean into the Atlantic Basin. This amazing amount of tectonic movement was due to Barbados being attached to the Caribbean Plate, which was moving rapidly between North and South America into the Atlantic Ocean. As the Caribbean Plate continued to move eastward, Barbados was pushed off to the side and eventually became stabilized on the Barbados Ridge east of Saint Vincent. Having been isolated from other landmasses and islands for so long, the deeper water, Bathyal Zone areas of Barbados have evolved a very rich and highly endemic molluscan fauna.

Of particular interest on the slopes of the Barbados Seamount is the presence of the cancellariid genus *Aphera*, which was thought to have become extinct within the Caribbean area during the Pliocene Epoch. The genus persists in the Panamic Province of the Eastern Pacific as a single species *Aphera tessellata*, but living Caribbean representatives were thought to be nonexistent. During a research trip to Barbados in 1979 (accompanied by M.G. Harasewych), I was given a large study collection of Barbados and Venezuelan deep-water shells collected by Dr. Finn Sander, the Director of the McGill University Marine Laboratory in St. James. In the collection was a specimen of a live-taken undescribed *Aphera* species, which was mislocalized as coming from Venezuela. The specimen was later found out to actually have come from 200 m depth off St. James, Barbados, showing that the only living Caribbean member of this "extinct" genus, *Aphera lindae* (Figure 9.7B), is a Barbados endemic gastropod. This unusual cancellariid was also found to occur along with several rarely seen southern Caribbean gastropods, including the muricids *Lindapterys sanderi* and *Poirieria hystricina* and the Lesser Antilles deep water cone shells *Sandericonus sanderi* and *Dalliconus mazei*. Some of the more prominent and important deep water gastropods of the Barbados Seamount include the following (several illustrated on Figure 9.7):

Muricidae-Ergalitaxinae
   Lindapterys sanderi (a subspecies, *rosalimae*, occurs in the Cearaian Subprovince)
Columbellidae
   Strombina (Cotonopsis) lindae (endemic to Barbados)
Olividae
   Oliva (Americoliva) barbadensis
Mitridae
   Mitra (Nebularia) lenhilli (endemic to Barbados)
Cancellariidae
   Aphera lindae (endemic to Barbados)
Conidae
   Sandericonus hunti
   Sandericonus knudseni
   Sandericonus sanderi

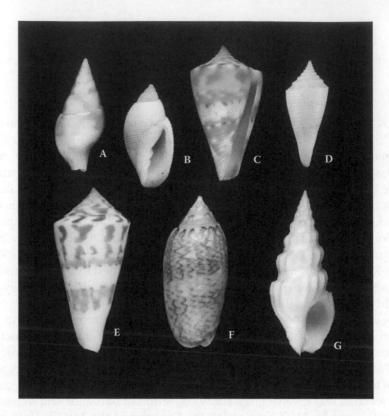

Figure 9.7 Gastropods of the Deep Water Areas of the Grenadian Subprovince: (A) *Strombina (Cotonopsis) lindae* (Petuch, 1988), length 21 mm. (B) *Aphera lindae* Petuch, 1987, holotype, length 10 mm. (C) *Sandericonus sanderi* (Wils and Moolenbeek, 1979), holotype, length 21.1 mm. (D) *Dalliconus kremerorum* (Petuch, 1988), holotype, length 18 mm. (E) *Sandericonus sorensoni* (Sander, 1982), length 56 mm. (F) *Oliva (Americoliva) barbadensis* Petuch and Sargent, 1986, length 43 mm. (G) *Clathrodrillia petuchi* (Tippett, 1995), length 53 mm.

> *Sandericonus sorensoni*
> *Tenorioconus cf. mappa* (probably a new species) (endemic to Barbados)
> Conilithidae
> *Dalliconus coletteae* (see Appendix 2)
> *Dalliconus kremerorum* (endemic to Barbados)
> Drilliidae
> *Clathrodrillia petuchi* (endemic to Barbados)
> Turridae
> *Polystira hilli* (endemic to Barbados)

The unusual richness and high endemicity of the Barbados bathyal malacofauna is a reflection of the long-term isolation that characterizes the geological history of the island. Of particular importance within the bathyal molluscan assemblages is the presence of a species radiation of the conid genus *Sandericonus*, containing at least four species (see Appendix 2). Although some workers have synonymized three of the four species with the taxon *Sandericonus sanderi* (Figure 9.7C), all four appear to represent valid species, especially the elongated *Sandericonus sorenseni* (Figure 9.7E). These deep water cones have also been collected in Martinique, Guadeloupe, and Curaçao.

Molluscan biodiversity in the Surinamian Subprovince

While the Grenadian Subprovince is composed entirely of island archipelagos, the Surinamian Subprovince is composed almost entirely of continental shorelines. Extending from the Orinoco River Delta and southern Trinidad south to the Territory of Amapá, Brazil and the Mouth of the Amazon River, the Surinamian Subprovince (Figure 9.8) encompasses vast areas of muddy mangrove-lined shores and offshore organic-rich muddy sea floors. These are the result of a tremendous amount of fresh water effluent eminating from hundreds of rivers flowing down from the Guiana Highlands of Guyana, Suriname, and French Guiana. With the Orinoco Delta in the north and the Amazon River Delta in the south, and with major rivers such as the Essequibo and Berbice of Guyana, the Paramaribo and Marowjine of Suriname, and the Oyapock of French Guiana, the entire subprovince is essentially one giant tropical estuary and has a very impoverished intertidal molluscan fauna. The ecologically less-stressed offshore areas have a richer fauna with higher percentages of endemism. Utilizing the raw data listed in Appendix 1 and based on the percentages of endemism in the 10 key index taxa, the Surinamian Subprovince can be seen to be a problematical area. Based on the relationship

$$t = \frac{n}{N}(100), \; t > 25$$

these are as follows:

Modulidae ($t1$), $t = 0$, with $N = 1$, $n = 0$
Turbinellidae ($t2$), $t = 50$, with $N = 2$, $n = 1$
Conidae ($t3$), $t = 33$, with $N = 6$, $n = 2$
Conilithidae ($t4$), $t = 20$, with $N = 5$, $n = 1$
Muricinae ($t5$), $t = 38$, with $N = 8$, $n = 3$
Fasciolariinae ($t6$), $t = 0$, with $N = 1$, $n = 0$
Lyriinae ($t7$), $t = 0$, with $N = 0$, $n = 0$ (absent from the Surinamian Subprovince)
Olivinae ($t8$), $t = 0$, with $N = 2$, $n = 0$
Cancellariinae ($t9$), $t = 0$, with $N = 2$, $n = 0$
Plesiotritoninae ($t10$), $t = 0$, with $N = 1$, $n = 0$

The Subprovincial Combined Index, incorporating the endemicity of these 10 families and subfamilies and defined by the relationship

$$S = \sum_{n=1}^{10} \frac{tn}{10}, \; S > 25$$

yields only the quantity S = 25. This relatively low level of subprovincial endemism (S>25) reflects the impoverishment of the the coastal molluscan faunas caused by fluctuating salinities and muddy substrates. If the most species-rich families (the Turbinellidae, Conidae, Conilithidae, and Muricinae) are factored by themselves, then S(t2, t3, t4, t5) = 35.3, a quantity large enough to support subprovincial status. Taking into account these 10 families and subfamilies, the Surinamian Subprovince is now known to house an interesting, but low diversity, malacofauna with a fairly high level of endemism. Some of these endemic mollusks include (several illustrated on Figure 9.9):

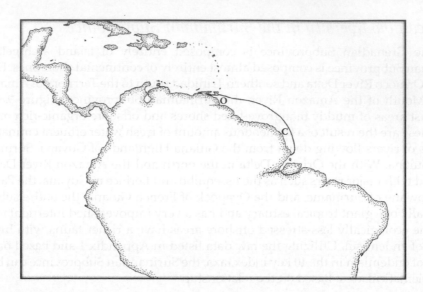

Figure 9.8 Map of the Surinamian Subprovince of the Caribbean Province, showing important geographical features. O = Orinoco River Mouth; C = Cabo Orange, Territory of Amapá, Brazil.

Gastropoda
Muricidae
   Favartia massemini
   Siratus lamyi
   Siratus springeri (especially common along the Amapá coast of Brazil)
   Siratus thompsoni (especially common along the Amapá coast of Brazil)
Fasciolariidae
   Fusinus benjamini (also found south of Barbados)
Turbinellidae
   Turbinella laevigata rianae (possibly a full, distinct species)
Olividae
   Oliva (Americoliva) new species (similar to the Brazilian *O. circinata*)
Conidae
   Conasprelloides bajanensis (also found south of Barbados)
   Conasprelloides brunneobandatus (also found off the Orinoco River Delta)
   Conasprelloides guyanensis
   Kellyconus rachelae (*Note*: this species was originally described as coming from the
      Golfo de Venezuela off Punto Fijo. This data, supplied by shrimp boat captains, is
      now known to be erroneous, and the holotype actually comes from 100 m depth
      off Boca Araguao, Orinoco River Delta, Venezuela)
   Sandericonus perprotractus (*Note*: this species was originally described as coming from
      the Golfo Triste of Venezuela. This data, supplied by shrimp boat captains, is now
      known to be erroneous, and the holotype actually comes from 100 m depth off
      Boca Araguao, Orinoco River Delta, Venezuela)
Bivalvia
Cuspidariidae
   Cardiomya surinamensis

Figure 9.9 Gastropods of the Surinamian Subprovince: (A) *Kellyconus rachelae* (Petuch, 1988), holotype, length 24 mm. (B) *Conasprelloides bajanensis* (Usticke, 1968), holotype, length 31 mm. (C) *Conasprelloides guyanensis* (Van Mol, 1973), holotype, length 28.4 mm. (D) *Conasprelloides brunneobandatus* (Petuch, 1992), holotype, length 28 mm (also found along the Venezuelan coast). (E) *Sandericonus perprotractus* (Petuch, 1987), holotype, length 46 mm. (F) *Vokesimurex donmoorei* (Bullis, 1964), length 51 mm (also found along Venezuela).

The widespread southern Caribbean olivellid *Olivella olssoni*, which was originally described from the muddy coastline of Suriname (Altena, 1971), also occurs along with these other endemic species. Very little malacological research has been conducted in the area of the Surinamian Subprovince, with Altena's paper and the work of Merle and Garrigues (2008; on new muricids) being some of the only publications. Okutani's 1982 paper on new species from Suriname is filled with synonyms and mislocalized taxa. His *"Dallivoluta surinamensis"* and *"Murex surinamensis,"* both supposedly from off the Suriname and Guyana coasts, are actually species that were collected in the Indian Ocean and are now known to be endemic to the Saya de Malha Banks south of the Seychelles Islands.

Amazonian faunal barrier

The coastline of the Territory of Amapá in northernmost Brazil is inundated by immense amounts of freshwater that is carried northward by the Guyana Current, a branch of the North Equatorial Current. The entire Amapá area, from Cabo Orange to Cabo Norte at the Mouth of the Amazon River, is almost entirely made up of brackish water mangrove

Figure 9.10 View of the Amapá coast of northern Brazil, showing one of the many distributaries of the Amazon River Delta south of Cabo Norte. Here, the brackish and fresh water mud flats and mangrove forests at the mouths of the distributaries can be seen to line the entire coastline. Physiologically hardy gastropods, capable of tolerating low salinities, are common on these mud flats and include *Pugilina morio*, *Thaisella trinitatensis*, *Neritina meleagris*, and *Neritina zebra*. (From E. Petuch, archival photograph, 1978.)

jungles and immense intertidal mud flats with low salinities (Figure 9.10). This impressive stretch of fresh and brackish coastal waters, over 700 km long, acts as an ecological barrier to most intertidal or subtidal mollusks from the Caribbean and Brazilian Provinces. These water conditions, in turn, often extend offshore for over 200 km, also acting as a barrier for deeper water, subtidal species. The Amazon Faunal Barrier was established during the early Pleistocene, when the older, Pliocene-aged Amazonian Sea marine environment became filled with sediments from the eroding Andes Mountains. By the mid-Pleistocene, the Amazonian Sea ceased to exist and was replaced completely by the sediment-filled Amazon River, which rapidly formed the Amazon Delta and its fresh water environments (Petuch, 1988; 2004). Once the ecological barrier was established, the isolated marine faunas on both sides began to undergo genetic drift and selective extinctions, leading to the great differences seen in the compositions of the malacofaunas of the Recent Caribbean and Brazilian Provinces.

Although being under the influence of variable amounts of fresh water, the Amapá and Amazon River Mouth areas do support an interesting fauna of euryhaline and eurytopic

Figure 9.11 Gastropods of the Amazon River Delta: (A) *Pugilina morio* (Linnaeus, 1758), length 128 mm. (B) *Thaisella trinitatensis* (Guppy, 1869), length 31 mm. (C) *Neritina zebra* (Bruguiere, 1792), length 15 mm. (D) *Neritina virginea braziliana* (Sowerby, 1849), length 12 mm. (E) *Neritina meleagris* Lamarck, 1822, length 13 mm. (F) *Neritina zebra* form *sobrina* Recluz, 1846, length 14 mm.

gastropods. Of principal interest here is a large fauna of the neritid gastropod genus *Neritina*, with at least five species occurring in great abundance on the brackish water mud flats and mangrove roots. The largest gastropod found on these mud flats is the amphiatlantic melongenid *Pugilina morio*, which also occurs all along West Africa, from Senegal to Angola. This voracious molluscivore ranges along a large area of the South American coastline, from Trinidad and the Orinoco River Mouth south to Espiritu Santo State, Brazil. Living with *Pugilina* are the euryhaline muricids *Thaisella trinitatensis* and *Thaisella coronata*, which are also voracious molluscivores that can be found feeding on *Crassostrea rhizophorae* mangrove oysters from Trinidad south to Espiritu Santo. Some of the more interesting and important brackish water gastropods from the area of the Amazonian Faunal Barrier include the following (several illustrated in Figure 9.11):

Neritidae
    Neritina cf. clenchi
    Neritina meleagris
    Neritina piratica
    Neritina virginea braziliana
    Neritina zebra
    Neritina zebra form *sobrina*

Littorinidae
  Littorinopsis angulifera
  Littorinopsis flava
Assimineidae
  Assiminea succinea
Muricidae
  Thaisella coronata (amphiatlantic; also found along West Africa)
  Thaisella trinitatensis
Melongenidae
  Pugilina morio (amphiatlantic; also found along West Africa)

None of these taxa are endemic to the subprovince, but they represent an interesting faunal overlap between the Brazilian and Caribbean Provinces. Future research in this mostly unexplored region of northernmost Brazil will probably uncover new endemic brackish water mollusks, most probably microgastropods.

chapter ten

Northern subprovinces of the Brazilian Province

East of the Amazon River mouth and the Baia de Marajó, along the coasts of the Pará and Maranhão States in Brazil, the oceanography of the neritic environments changes abruptly. Instead of being dominated by organic-rich mud, the sea floors along this coast are composed of carbonate sand mixed with quartz sand and typically contain large bioherms of calcareous red coralline algae. The northward-flowing North Equatorial Current, which deflects northward at the "nose" of Brazil near Recife, Pernambuco State, brings warm, eutropical conditions to this area and pushes the Amazonian muds away and up the coast to Amapá and French Guiana. South of the nose of Brazil, the southward-flowing Brazilian Current carries warm tropical water along the coasts of Bahia and Espiritu Santo States as far south as Cabo Frio, Rio de Janeiro State and creates warm-temperate water conditions as far south as the Rio de la Plata in Argentina. These two currents essentially divide northern and central Brazil into two distinct subprovinces of the Brazilian province: the Cearaian (named for Ceará State, Brazil); and the Bahian (named for Bahia State, Brazil) (Figure 10.1). Although they have separate, distinct malacofaunas, the Cearaian and Bahian Subprovinces also share many Brazilian endemic species, producing a small amount of faunal overlap.

Molluscan biodiversity in the Cearaian Subprovince

Extending from Cabo Orange, Amapá Territory to approximately Maceió, Alagoas State, the Cearaian Subprovince encompasses the entire northern coastline of Brazil (Figure 10.1). Also included within the subprovince are two small island groups off the coast of Rio Grande do Norte State: the Atol das Rocas and Fernando de Noronha. Characteristically, the shoreline areas of northern Brazil are predominantly sandy, with wide beaches and giant dunes, with some, like the immense seif dunes of coastal Ceará State, being among the largest in South America. The offshore areas are typically composed of carbonate environments, with coarse shell hash and coral sands, coralline algal bioherms, and small coral reefs. Along the shorelines from Acarau and Fortaleza, Ceará State, south to Maceió, Alagoas State, the red coralline algae *Lithothamnion, Mesophyllum, Sporolithon*, and *Lithophyllum* form massive fused rhodolithic algal ridges and "reefs" that follow the contours of the entire coast. The combined molluscan fauna of the quiet, protected sandy bays; the offshore coralline algae rhodolith beds and carbonate sand; and the massive red algal coastal ridges has proven to be highly endemic, setting aside the Cearaian Subprovince from other areas in the Brazilian Province.

Of special interest in this northernmost Brazilian Subprovince are several endemic genera and species complexes, all restricted to the areas within the subprovincial boundaries. One of these, the conilithid genus *Artemidiconus*, comprising at least two species, is found only on the carbonate sea floors off the northern Brazilian coast, from Pará to Sergipe. The rare endemic lucinid bivalve genus *Miltha*, which occurs in the Eastern Pacific Panamic Province and is a relict from the early Pleistocene, also lives on these offshore

Figure 10.1 Map of the Brazilian Province, showing its four subprovinces. The northern subprovinces, the Cearaian (1) and the Bahian (2), can be seen to encompass more than half of the Brazilian coastline. Important geographical features in the Cearaian and Bahian Subprovinces include the following: AM = Amazon River Mouth; R = Atol das Rocas; N = Fernando de Noronha Island; S = Salvador and Todos os Santos Bay, Bahia State; A = Abrolhos Archipelago; and T = Trindade Island. Farther south, in the Paulinian Subprovince (3) and Uruguayan Provinciatone (4), the important geographical features include the following: RJ = Rio de Janeiro, Rio de Janeiro State; and F = Florianopolis, Santa Catarina State.

carbonate areas, from Pará to northern Bahia State. The large endemic volute, *Voluta ebraea*, has the same range as *Miltha* and is a prominent component of a singular tropical molluscan assemblage unlike any other in the western Atlantic. Using the raw data listed in Appendix 1 and based on the percentages of endemism in the 10 key index taxa, the Cearaian Subprovince can be seen to be a discrete biogeographical entity within the Brazilian Province. Applying the relationship

$$t = \frac{n}{N}(100),\ t > 25$$

these are as follows:

Modulidae (*t*1), *t* = 50, with *N* = 2, *n* = 1
Turbinellidae (*t*2), *t* = 0, with *N* = 2, *n* = 0

Conidae (*t*3), *t* = 60, with *N* = 5, *n* = 3
Conilithidae (*t*4), *t* = 86, with *N* = 7, *n* = 6
Muricinae (*t*5), *t* = 17, *N* = 12, *n* = 2
Fasciolariinae (*t*6), *t* = 33, *N* = 3, *n* = 1
Lyriinae (*t*7), *t* = 100, *N* = 1, *n* = 1
Olivinae (*t*8), *t* = 50, *N* = 2, *n* = 1
Cancellariinae (*t*9), *t* = 66, *N* = 3, *n* = 2
Plesiotritoninae (*t*10), *t* = 0, *N* = 1, *n* = 0

The subprovincial combined index, incorporating the endemicity of these 10 families and subfamilies and defined by the relationship

$$S = \sum_{n=1}^{10} \frac{tn}{10}, \, S > 25$$

yields the quantity $S = 46.2$. This extremely high level of endemicity ($S > 25$), approaching full provincial status, demonstrates that the Cearaian Subprovince is a strong biogeographical entity. Some of the more important and prominent endemic mollusks of the Cearaian Subprovince include the following (several shown in Figures 10.2 and 10.3):

Gastropoda
Cerithiidae
  Bayericerithium bayeri
Muricidae
  Favartia coltrorum
  Lindapterys sanderi rosalimae
  Poirieria oregonia
Buccinidae
  Caducifer atlanticus
Volutidae
  Voluta ebraea
Harpidae-Moruminae
  Cancellomorum matthewsi
Volutidae
  Voluta ebraea
Olividae
  Ancilla faustoi
  Eburna lienardi
  Eburna matthewsi
  Oliva (Americoliva) circinata jorioi (see Appendix 2)
Marginellidae
  Bullata lilacina
  Bullata matthewsi
Conidae
  Brasiliconus scopolorum
  Poremskiconus mauricioi

Figure 10.2 Endemic gastropods of the Cearaian Subprovince: (A) *Ancilla faustoi* Matthews and Matthews, 1978, length 12 mm. (B) *Voluta ebraea* Linnaeus, 1758, length 124 mm. (C) *Bayericerithium bayeri* Petuch, 2001, length 25 mm. (D) *Bullata lilacina* (Sowerby, 1846), length 27 mm. (E) *Artemidiconus selenae* (Van Mol, Tursch, and Kempf, 1967), length 11 mm. (F) *Favartia coltrorum* Houart, 2005, length 10 mm.

> Conilithidae
> > *Artemidiconus selenae*
> > *Artemidiconus yemanjae*
> > *Jaspidiconus damasoi*
> Bivalvia
> Lucinidae
> > *Miltha childrenae*

The coralline carbonate sea floors and rhodolith beds of northeastern Brazil are unique in South America, and this special substrate type is reflected in the rich and highly endemic malacofauna. The algal shoreline ridges are best developed along the stretch of coastline between Natal and Maceió and produce the "reef" for which the city of Recife is named.

Atol das Rocas and Fernando de Noronha Island

The Brazilian island groups of the Atol das Rocas and Fernando de Noronha together constitute the territory of Fernando de Noronha. The Atol das Rocas, one of only 13 true coral atolls in the western Atlantic, lies 145 km west of Fernando de Noronha and 260 km

Figure 10.3 Endemic gastropods of the Cearaian Subprovince: (A) *Artemidiconus yemanjae* (Van Mol, Tursch, and Kempf, 1967), holotype, length 12 mm. (B) *Brasiliconus scopolorum* (Van Mol, Tursch, and Kempf, 1967), holotype, length 20.7 mm. (C) *Caducifer atlanticus* Coelho, Matthews, and Cardoso, 1970, length 16 mm. (D) *Cancellomorum matthewsi* (Emerson, 1967), length 25 mm. (E) *Poremskiconus mauricioi* (Coltro, 2004), length 21 mm. (F) *Oliva (Americoliva) circinata jorioi* Petuch, new subspecies, holotype, length 32 mm (see Appendix 2).

off the coast of Natal. Fernando de Noronha, a group of 21 small, high volcanic islands, lies 545 km off Pernambuco and is the farthest of the two island groups from the Brazilian mainland. Both island groups harbor extensive coral reef complexes and offer a wide variety of habitats for shallow-water mollusks. Because of this large variety of substrate types and because of the extreme geographical and oceanographic isolation, the Atol das Rocas and Fernando de Noronha together have produced a number of interesting endemic species. On the 21 islands of the Fernando de Noronha Group, the greater part of the gastropod fauna of the rocky intertidal zone is composed of endemic species, including members of the genera *Polygona* (Petuch, 1986b), *Lottia, Thais, Echinolittorina*, and *Nerita*. Several other shallow-water endemic gastropods are found in the sandy pockets between the rock outcrops and in exposed tide pools. Some of these Fernando de Noronha intertidal endemics include the following (with three shown in Figure 10.4):

Lottiidae
    Lottia noronhensis
Littorinidae
    Echinolittorina vermeiji

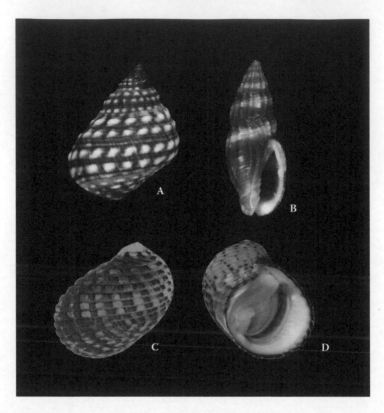

Figure 10.4 Endemic gastropods from Fernando de Noronha Island: (A) *Echinolittorina vermeiji* (Bandel and Kadolsky, 1982), length 9 mm. (B) *Enaeta leonardhilli* Petuch, 1988, length 12 mm. (C, D) *Nerita ascensionis deturpensis* Vermeij, 1970, width 34 mm.

Neritidae
    Nerita ascensionis deturpensis
Muricidae
    Thais meretricula
Fasciolariidae
    Polygona vermeiji
Volutidae
    Enaeta leonardhilli

Living in deeper water around the reefs of Atol das Rocas and Fernando de Noronha is the endemic Grinning Tun, *Malea (Quimalea) noronhensis* (Tonnidae), the only living member of its genus and subgenus to occur in the Atlantic Ocean. The subgenus *Quimalea* is common on coral reefs in the Indo-Pacific region, and its anomalous occurrence on a few tiny islands off northern Brazil is problematic. The Fernando de Noronha population of the western Atlantic amphiprovincial conid *Stephanoconus regius* differs from typical specimens in being much darker and heavier and resembling the Panamic *Stephanoconus brunneus*, and it may represent a new subspecies. More research regarding these two island groups will doubtlessly uncover many new endemic species.

Molluscan biodiversity in the Bahian Subprovince

Extending from Maceió, Alagoas State south to Cabo Frio, Rio de Janeiro State, and encompassing the offshore Abrolhos Archipelago and reef complexes and remote Trindade Island, the Bahian Subprovince is spatially the largest of the biogeographical subdivisions of the Brazilian Province. The Bahian Subprovince also contains the most species-rich malacofauna in Brazil and one of the highest levels of endemicity found in any subprovince in the tropical western Atlantic region. Several gastropod genera are endemic to the subprovince, with the volutid genus *Plicoliva* being one of the more important examples. Research by Philippe Bouchet of the Paris Museum of Natural History showed that not only was this unusual little Bahian volutid a new genus but also that it actually represented a previously unknown subfamily of the Volutidae (now known as the Plicolivinae, with only three members—two in Brazil and one in West Africa). The subprovince also contains large endemic radiations of the conid genus *Poremskiconus* (with at least six species) and the Bahian endemic conolithid genus *Coltroconus* (with at least three species).

By using the raw data listed in Appendix 1 and by calculating the percentages of endemism in the 10 key index taxa, the Bahian Subprovince can be seen to be a differentiable subdivision of the Brazilian Province. Applying the relationship

$$t = \frac{n}{N}(100), \; t > 25$$

these are as follows:

Modulidae (*t*1), $t = 0$, with $N = 1$, $n = 0$
Turbinellidae (*t*2), $t = 0$, with $N = 2$, $n = 0$
Conidae (*t*3), $t = 83$, with $N = 12$, $n = 10$
Conilithidae (*t*4), $t = 86$, with $N = 7$, $n = 6$
Muricinae (*t*5), $t = 50$, with $N = 10$, $n = 5$
Fasciolariinae (*t*6), $t = 33$, with $N = 3$, $n = 1$
Lyriinae (*t*7), $t = 100$, with $N = 1$, $n = 1$
Olivinae (*t*8), $t = 0$, with $N = 1$, $n = 0$
Cancellariinae (*t*9), $t = 33$, with $N = 3$, $n = 1$
Plesiotritoninae (*t*10), $t = 100$, with $N = 1$, $n = 1$

The subprovincial combined index, incorporating the endemicity of these 10 families and subfamilies and defined by the relationship

$$S = \sum_{n=1}^{10} \frac{tn}{10}, \; S > 25$$

yields the quantity $S = 61$. This extremely high level of endemism ($S > 25$), which approximates full provincial status, readily demonstrates that the Bahian Subprovince is a strong biogeographical subdivision within the Brazilian Province and that it is the center of speciation for the entire province.

Figure 10.5 Widespread gastropods of the Cearaian and Bahian Subprovinces: (A) *Pisania janei-rensis* (Philippi, 1848), length 28 mm. (B) *Siratus coltrorum* (E. Vokes, 1990), length 32 mm. (C) *Bursa ponderosa* (Reeve, 1844), length 43 mm. (D) *Haliotis aurantium* Simone, 1998, length 11 mm. (E, F) *Luria cinerea brasiliana* Lorenz and Hubert, 1993, length 23 mm. (G) *Polystira coltrorum* Petuch, 1993, length 79 mm.

Besides the widespread Brazilian Province endemic taxa discussed and illustrated in Chapter 2 (see Figure 2.7), a number of other characteristic endemic taxa also occur throughout both the Cearaian and Bahian Subprovinces. Some of these widespread gastropods (Petuch, 1993b), which range from at least Pará State in the north all the way south to Cabo São Tome and Cabo Frio in northern Rio de Janeiro State, include the following (several illustrated in Figure 10.5):

Haliotidae
   Haliotis aurantium
Strombidae
   Titanostrombus goliath
Cypraeidae
   Luria cinerea brasiliensis
Bursidae
   Bursa ponderosa
Muricidae
   Siratus coltrorum

Figure 10.6 View of the coastline of southern Bahia State, Brazil, north of the city of Porto Seguro. Here, a rhodolithic coralline algal ridge lies just offshore and acts as a barrier between the brackish water coastal lagoon (filled with tannin-stained brown water) and the offshore neritic environments (blue-green open oceanic sea water). These algal "reefs," composed of the red corallines *Lithothamnion*, *Sporolithon*, *Mesophyllum*, and *Lithophyllum*, are typical of the coastal areas of northern and central Brazil, from Ceará State to southern Bahia State. (From E. Petuch, archival photograph, 1978.)

Fasciolariidae
  Leucozonia brasiliana
Buccinidae
  Pisania janeirensis
Olividae
  Oliva (Americoliva) circinata
Conilithidae
  Jaspidiconus pusillus
Turridae
  Polystira coltrorum

As in the Cearaian Subprovince, coralline algal reefs and rhodolith beds dominate the coastline of the Bahian Subprovince, from Salvador south to Ilheus, and often form wide coastal lagoons (Figure 10.6). These near-shore algal reefs and offshore rhodolith-based carbonate environments house a very large number of endemic mollusks, many with very restricted ranges. The coralline algae and scleractinian coral bioherms surrounding Itaparica Island in Todos os Santos Bay in Salvador have been found to be especially rich in endemic mollusks. Several of these, including the cone shell, *Jaspidiconus henckesi* (Conilithidae), the fasciolariid *Polygona bayeri*, and the muricid *Murexiella leonardhilli*, are restricted to the Todos os Santos Bay area, and the Salvador coast is now known to be a center of localized speciation. Some of the more important and prominent shallow water

Figure 10.7 Endemic cones shells of the Bahian Subprovince: (A) *Poremskiconus bertarollae* (Costa and Simone, 1997), length 21 mm. (B, C) *Coltroconus delucai* (Coltro, 2004), length 11 mm. (D) *Poremskiconus brasiliensis* (Clench, 1942), length 20 mm. (E) *Dauciconus worki* (Petuch, 1986), length 41 mm. (F) *Dauciconus riosi* (Petuch, 1986), length 49 mm.

Bahian endemic gastropods include (with several illustrated on Figures 10.7, 10.8, 10.9, and 10.10):

Turbinidae
  Turbo heisei
Modulidae
  Modulus bayeri
Turritellidae
  Torculoidella hookeri (also in the northern part of the Paulinian Subprovince)
Triviidae
  Dolichupis virgo
  Niveria brasilica
Cypraeidae
  Macrocypraea (Macrocypraea) zebra dissimilis (also in the northern part of the Paulinian Subprovince; may prove to be a full species)
Ovulidae
  Cyphoma macumba
Muricidae
  Dermomurex oxum
  Favartia varimutabilis

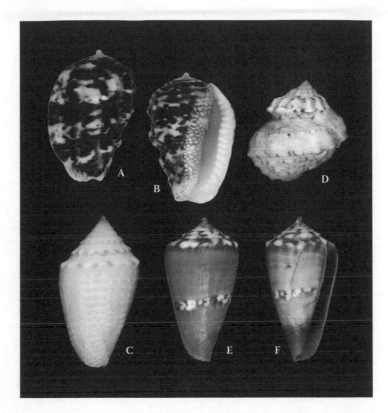

Figure 10.8 Endemic gastropods of the Bahian Subprovince: (A, B) *Morum bayeri* Petuch, 2001, length 23 mm. (C) *Jaspidiconus henckesi* (Coltro, 2004), length 13 mm. (D) *Modulus bayeri* Petuch, 2001, length 12 mm. (E, F) *Magelliconus pseudocardinalis* (Coltro, 2004), length 19 mm.

    Murexiella iemanja
    Murexiella leonardhilli
    Muricopsis josei
    Pygmaepterys oxossi
Fasciolariidae
    Polygona bayeri
    Polygona ogum
Buccinidae
    Pisania bernardoi
Volutidae
    Plicoliva zelindae
Mitridae
    Mitra brasiliensis
Harpidae-Moruminae
    Morum bayeri
Marginellidae
    Bullata bullata
Olividae
    Oliva (Americoliva) circinata
    Olivancillaria steeriae

Figure 10.9 Endemic gastropods of the Bahian Subprovince: (A) *Plicoliva zelindae* Petuch, 1979, length 29 mm. (B) *Pisania bernardoi* Costa and Gomes, 1998, length 11 mm. (C) *Tritonoharpa leali* Harasewych, Petit, and Verhecken, 1992, length 17 mm. (D) *Oliva (Americoliva) circinata* Marrat, 1870, length 58 mm. (E) *Murexiella iemanja* Petuch, 1979, length 7 mm.

    Conidae
        Dauciconus riosi
        Dauciconus worki
        Magelliconus pseudocardinalis (the only member of its genus in Brazil)
        Poremskiconus abrolhosensis (= *baiano*)
        Poremskiconus archetypus
        Poremskiconus bertarollae
        Poremskiconus brasiliensis
        Poremskiconus cargilei
    Conilithidae
        Coltroconus delucai
        Coltroconus iansa (= *bodarti*)
        Coltroconus schirrmeisteri
        Jaspidiconus cf. duvali (probably a new species)
        Jaspidiconus henckesi

    The Bahian Subprovince endemic genus *Coltroconus*, encompassing *delucai*, *iansa*, and *schirrmeisteri*, exhibits large knobs along the shoulders, a character not seen in other conilithids found elsewhere in the western Atlantic. These heavily knobbed Bahian endemic conilithids, which resemble miniature *Stephanoconus* species, appear to represent

Figure 10.10 Endemic gastropods of the Bahian Subprovince: (A) *Muricopsis josei* E. Vokes, 1994, length 14 mm. (B) *Mitra brasiliensis* Oliveira, 1969, length 19 mm. (C) *Polygona ogum* (Petuch, 1979), length 38 mm. (D) *Pygmaepterys oxossi* (Petuch, 1979), length 10 mm. (E) *Turbo (Taenioturbo) heisei* Prado, 1999. (F) *Murexiella leonardhilli* Petuch, 1987, length 36 mm.

a localized radiation that is centered on the Abrolhos Archipelago and reef complexes (Coltro, 2011).

Abrolhos Archipelago and reef complexes

Located 65 km off the coast of Caravelas, southern Bahia State, the Abrolhos Archipelago comprises five small volcanic islands (Figure 10.11) and numerous large coral reef complexes, including the Parcel das Paredes, Timbebas, and Pedra Lixa reefs. The scleractinian coral complexes of this large platform represent the southernmost major coral reef growth in the western Atlantic. Because of their extreme southerly geographical position and marginally tropical water temperatures, the reefs of the Abrolhos Archipelago are made up of only a few species of hardy corals, with most of them being endemic to the southern Brazilian coast. On the Timbebas Reefs and Parcel das Paredes, the largest of the reef complexes, the scleractinians *Mussismilia harttii* and *Mussismilia braziliensis* make up the greater part of the coral fauna. The massive growth forms of these two corals form the substrate for a fascinating fauna of endemic invertebrates, including several species of large comatulid crinoids (feather stars) and the oak leaf–shaped gorgonian *Phyllogorgia dilatata*. Living along with these endemic cnidarians and echinoderms is a large fauna of gastropod mollusks, many of which are endemic to the Abrolhos Islands and reefs and adjacent coast.

Figure 10.11 View of Santa Barbara Island, the largest of the Abrolhos Islands, Bahia State, Brazil. The lighthouse and a weather station are manned by several Brazilian sailors, and the small settlement houses their families. The endemic limpet, *Lottia abrolhosensis*, is found on the black basaltic boulders at the base of the steep cliffs. (From E. Petuch, archival photograph, 1978.)

I had the wonderful opportunity to visit the then-unexplored Abrolhos Islands in 1977, as a diving and collecting partner for a Brazilian doctoral student at the University of Miami (Dr. Zelinda Leão, now of the University of Salvador). Previous to my research trip, the islands had been extensively studied by only two other biologists: Charles Darwin in 1832, while on the voyage of the *HMS Beagle*, and Jacques Laborel, who in 1962 undertook a detailed study of the Abrolhos corals while working on the research vessel *Calypso* with Jacques Cousteau. I consider it a great privilege to have been the first malacologist to study the Abrolhos molluscan fauna in any detail, and my intensive three weeks of fieldwork there resulted in the descriptions of 10 new species and two new genera (Petuch, 1979). Subsequent research on material collected on the reef complexes led to the description of three more new species. During my collecting on the Timbebas Reefs and neighboring Parcel das Paredes, I found living specimens of the fasciolariid *Polygona ogum*, the volutid *Plicoliva zelindae*, and the miters *Nodicostellaria kaicherae* and *Nodicostellaria lixa* crawling in shallow pools at low tide. It was absolutely amazing to find this many unnamed species sitting right out in the open in only millimeters of water. Upon visiting the largest island of the archipelago, Santa Barbara Island, I noticed that the common limpet that covered the smooth basaltic boulders at low tide by the thousands was also a new species that had never been described (which I later named *Lottia abrolhosensis*). After only a few days of observation and collecting in the Abrolhos, it was obvious that the area contained a highly endemic malacofauna with many unnamed taxa, and I consider this trip to be one of the most exciting field experiences that I have ever undertaken.

Figure 10.12 Endemic gastropods of the Bahian Subprovince: (A) *Plicoliva zelindae oceanica* Coltro, 2005, length 17 mm. (B) *Olivancillaria steeriae* Reeve, 1850, length 30 mm. (C) *Bullata bullata* (Born, 1778), length 62 mm. (D, E) *Macrocypraea (Macrocypraea) zebra dissimilis* Schilder, 1924, length 86 mm. (F) *Cyphoma macumba* Petuch, 1979, length 24 mm.

Some of the more important and prominent endemic gastropods of the Abrolhos Archipelago include the following (with several illustrated in Figure 10.12):

Lottiidae
    Lottia abrolhosensis
Volutidae
    Plicoliva zelindae oceanica (endemic to remote banks off the Abrolhos Islands)
Costellariidae
    Nodicostellaria kaicherae
    Nodicostellaria lixa
Conidae
    Poremskiconus abrolhosensis (= baiano)
Conilithidae
    Coltroconus iansa (= bodarti; most commonly found around the Abrolhos reef complexes)

Several of the species that I discovered and named during my research trip to the Abrolhos Archipelago (see Figure 10.13), such as the muricids *Pygmaepterys oxossi, Dermomurex oxum,* and *Murexiella iemanja* and the fasciolariid *Polygona ogum,* were later found to be living along the adjacent Bahia State mainland (which removes them from Abrolhos endemic status). All of these species, along with the Abrolhos conilithid *Coltroconus iansa,* were named in honor of the gods (Orixás) of the local Macumba–Candomblé religion, a West African–based faith brought to Bahia by Yoruba slaves.

Endemism on Trindade Island

The isolated Brazilian islands of Trindade and neighboring Martim Vaz are over 1,200 km off the coast of Espiritu Santo State and are the most remote of the Brazilian coastal islands. Trindade is the largest of the island group and lies 49 km from the four small islets that make up the Martim Vaz Islands. Although Trinidade houses an impoverished Bahian Subprovince malacofauna of only around 150 species, its geographical remoteness has produced genetic isolation in nonvagile groups and has led to the evolution of several endemic gastropods. As in the case of Atol das Rocas and Fernando de Noronha Island, the high intertidal zone of Trindade Island is dominated by several endemic species of the gastropod genera *Echinolittorina* (Littorinidae), *Nerita* (Neritidae), *Lottia* (Lottiinae), and *Leucozonia* (Fasciolariidae). Some of the more important endemic gastropods from this extreme easternmost edge of the Bahian Subprovince are as follows (with three illustrated in Figure 10.13):

Figure 10.13 Endemic gastropods of the Abrolhos Islands: (A) *Poremskiconus abrolhosensis* (Petuch, 1986), length 21 mm (the holotype of *abrolhosensis* is a juvenile specimen of what was later named *P. baiano* Coltro, 2004). (B, C) *Coltroconus iansa* (Petuch, 1979), length 13 mm. (= *bodarti* Coltro, 2004). (D) *Nodicostellaria lixa* Petuch, 1979, length 13 mm. (E) *Lottia abrolhosensis* (Petuch, 1979), length 23 mm. (F) *Nodicostellaria kaicherae* Petuch, 1979, length 8 mm.

Lottiidae
   Lottia marcusi
Turbinidae-Liotiinae
   Arene boucheti
Neritidae
   Nerita ascensionis trindadeensis
Littorinidae
   Echinolittorina vermeiji
Fasciolariidae
   Leucozonia ponderosa
Conidae
   Dauciconus jorioi (see Appendix 2)

A new species of the conid genus *Dauciconus* was recently discovered on the reefs off Trindade Island (shown in Figure 10.14A,B). This impressive endemic cone shell, *Dauciconus jorioi* (see Figure 10.14A), one of the largest conid species found in the Trindade-Martim Vaz Islands, appears to belong to the *Dauciconus riosi-Dauciconus worki* species complex of the mainland Bahian Subprovince. This large new species' presence on Trindade demonstrates the Bahian faunal affinities of this remote archipelago.

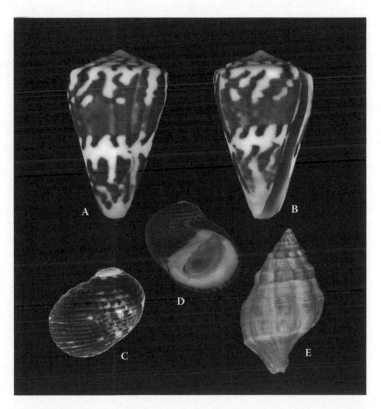

Figure 10.14 Endemic gastropods of Trindade Island: (A, B) *Dauciconus jorioi* Petuch, new species, holotype, length 53 mm (see Appendix 2). (C, D) *Nerita ascensionis trindadeensis* Vermeij, 1970, width 23 mm. (E) *Leucozonia ponderosa* Vermeij and Snyder, 1998, length 38 mm (smooth variant without shoulder knobs; may represent a new species).

chapter eleven

Molluscan biodiversity in the Paulinian Subprovince

The southernmost subprovince of the Brazilian Province, the Paulinian, encompasses the northern half of the Southwestern Atlantic Shelf, a physiographic region that extends from the Cabo São Tome–Cabo Frio area of Rio de Janeiro State south to the Falkland Islands off Argentina. The Paulinian Subprovince is centered around the South Brazil Bight, a wide, broad, sheltered embayment that extends from Rio de Janeiro, Rio de Janeiro State (Figure 11.1) south to Florianopolis, Santa Catarina State, and contains complex marine environments composed of small coral and sponge bioherms and coralline algal ridges and rhodolith beds. Oceanographically, the South Brazil Bight is completely within the influence of the warm Brazilian Current, which supports a eutropical molluscan fauna complete with almost all the tropical index taxa. South of Florianopolis on the Southwestern Atlantic Shelf, the Brazilian Current swings farther offshore and cool water eddies spinning off the subantarctic Falklands Current migrate up and down the coast (Matano, Palma, and Piola, 2010), creating only marginally paratropical conditions. South of Rio Grande, Rio Grande do Sul State, the northward-flowing eddies off the Falklands Current become stronger and more dominant, and nine of the ten tropical index taxa groups disappear. Only one cold-tolerant species of conid, *Lamniconus carcellesi*, occurs in this area and ranges as far south as the mouth of the Rio de la Plata in Uruguay and Argentina. This southernmost area of Brazilian influence is here referred to as the Uruguayan Provinciatone, and the area is a broad transition zone between the tropical Brazilian Province and the subantarctic Patagonian Province.

Extending from Cabo São Tome and Macae, in the Cabo Frio area of Rio de Janeiro State, south to Rio Grande, Rio Grande do Sul State, the Paulinian Subprovince (named for São Paulo State) contains the southernmost tropical malacofauna in the western Atlantic. Members of 9 of the 10 index families and subfamilies are present within the subprovincial boundaries and the molluscan fauna, in general, exhibits a high level of endemism. Two centers of localized speciation are present within the subprovince: one along the stretch of coast between Cabo São Tome and Cabo Frio, northern Rio de Janeiro State, and the area just north of the city of Rio de Janeiro; and one within the South Brazilian bight and its island groups (off southern Rio de Janeiro, São Paulo, and Santa Catarina States). Of special interest within the Paulinian Subprovince is a large species radiation of the endemic conid genus *Lamniconus* (Petuch, 1986) and the volutid genus *Odontocymbiola* (Figure 11.2; with at least five species).

By using the raw data listed in Appendix 1 and by calculating the percentages of endemism in the 10 key index taxa, the Paulinian Subprovince can be seen to be a distinct biogeographical subdivision of the Brazilian Province. Applying the relationship

$$t = \frac{n}{N}(100), \, t > 25$$

Figure 11.1 View of Rio de Janeiro and the adjacent coastline from the top of Mount Corcavado. Here, the large central lagoon, Lagoa Rodrigo de Freitas, can be seen to separate the Humaita section of Rio from the beach communities of Leblon and Ipanema. The Cagarras Islands can be seen 5 km offshore of Ipanema Beach. Because of the massive upwelling systems of the nearby Cabo Frio region, the waters off Rio de Janeiro and the Cagarras Islands remain cool all year, creating paratropical conditions. (From E. Petuch, archival photograph, 1978.)

these are as follows:

Modulidae ($t1$), $t = 0$, with $N = 1$, $n = 0$
Turbinellidae ($t2$), $t = 0$, with $N = 0$, $n = 0$ (absent from the Paulinian Subprovince)
Conidae ($t3$), $t = 100$, with $N = 7$, $n = 7$
Conilithidae ($t4$), $t = 100$, with $N = 2$, $n = 2$
Muricinae ($t5$), $t = 0$, with $N = 3$, $n = 0$
Fasciolariinae ($t6$), $t = 100$, with $N = 1$, $n = 1$
Lyriinae ($t7$), $t = 100$, with $N = 1$, $n = 1$
Olivinae ($t8$), $t = 100$, with $N = 1$, $n = 1$
Cancellariinae ($t9$), $t = 0$, with $N = 2$, $n = 0$
Plesiotritoninae ($t10$), $t = 0$, with $N = 1$, $n = 0$

The subprovincial combined index, incorporating the endemicity in these 10 families and subfamilies and defined by the relationship

Figure 11.2 Endemic gastropods of the Paulinian Subprovince and Cabo Frio region: (A) *Odontocymbiola americana* (Reeve, 1856), length 35 mm. (B) *Odontocymbiola simulatrix* Leal and Bouchet, 1989, length 72 mm. (C) *Odontocymbiola saotomensis* Calvo and Coltro, 1997, length 51 mm. (D) *Odontocymbiola macaensis* Calvo and Coltro, 1997, length 46 mm. (E) *Odontocymbiola cleryana* (Petit, 1856), length 47 mm. (F) *Cyphoma seccato* (Sowerby, 1834), length 19 mm.

$$S = \sum_{n=1}^{10} \frac{tn}{10}, \ S > 25$$

yields the quantity $S = 50$. This extremely high level of subprovincial endemism ($S > 25$), equal to a full province, readily demonstrates that the Paulinian Subprovince is a strong biogeographical entity with the highest level of endemicity seen in any subprovince of the tropical western Atlantic provinces.

Endemism in the Cabo Frio region

The area from Cabo São Tome and Cabo Frio to the area just north of the city of Rio de Janeiro, collectively referred to as the *Cabo Frio Region*, is now known to be a special center of molluscan evolution within the Paulinian Subprovince. Along this geographically small locality of northern Rio de Janeiro State, immense permanent wind-driven upwelling systems have developed, bathing the coastal and neritic areas with cool, nutrient-rich water (Gonzalez-Rodriguez et al., 1991; Castelao and Barth, 2006). The cold water of the "Cold Cape" (Cabo Frio) is produced by these upwellings, and the high nutrient level supports a large commercial shrimping industry. Most of the deeper water, offshore mollusks are collected by these shrimpers, and the area oceanographically resembles a miniature version of the upwelling systems off the Goajira Peninsula and Golfo de Venezuela. These constant high productivity oceanographic conditions have allowed for the evolution of unique marine communities and a distinctive local malacofauna. The "island" of cool water in the Cabo Frio region, bracketed between the warmer coastal waters of southern Espiritu Santo and southern Rio de Janeiro States, supports a number of distinctive endemic gastropods, including the following (several of which are shown in Figure 11.2):

Triviidae
  Pusula macaeica
Ovulidae
  Cyphoma versicolor
  Pseudocyphoma rosenbergi
Columbellidae
  Mitrella cabofrioensis
Volutidae
  Nanomelon viperinum
  Odontocymbiola cleryana
  Odontocymbiola saotomensis
Conidae
  Sandericonus carioca
  Sandericonus ednae
  Lamniconus clenchi
  Lamniconus xanthocinctus
Terebridae
  Strioterebrum brasiliensis

With the exception of the highly localized volute *Odontocymbiola saotomensis* and the ovulid *Cyphoma versicolor*, all the other species extend slightly northward into the extreme southern end of the adjacent Espiritu Santo State, where they occur together with a complete compliment of classic Bahian Subprovince gastropods. These *Pusula*, *Pseudocyphoma*, *Odontocymbiola*, and *Lamniconus* species do not occur south of Cabo Frio, where they are replaced by an endemic South Brazil Bight fauna of congeners. Recent intensive collecting in this area by noted divers and marine specimen collectors Afonso Jorio and Luiz Couto (both of Guaraparí, Espiritu Santo) and Jose and Marcus Coltro (both of São Paulo) have brought to light many more new, undescribed species, demonstrating that the Cabo Frio region has a much larger endemic fauna than was originally thought.

Endemism in the South Brazil Bight

The core of the Paulinian Subprovince, the large embayment of the South Brazil Bight, contains a distinctive and geographically restricted molluscan fauna. Although subject to cooler water temperatures during the winter months, the embayment retains a constant mild temperature year-round, allowing for the establishment of tropically appearing and

Figure 11.3 Endemic gastropods of the Paulinian Subprovince: (A) *Bullata analuciae* Souza and Coovert, 2001, length 32 mm. (B) *Agaronia travassosi* Morretes, 1938, length 52 mm. (C) *Cancellaria petuchi* Harasewych, Petit, and Verhecken, 1992, length 38 mm. (D) *Coronium elegans* Simone, 1996, length 55 mm. (E) *Oliva (Americoliva) circinata tostesi* Petuch, 1987, length 51 mm. (F) *Charonia lampas pustulata* (Euthyme, 1889), length 114 mm.

tropically derived molluscan assemblages. Similar environments containing rich faunas of tropically derived taxa are seen in the Cape Province of South Africa and the Great Australian Bight of southern Australia. As in the case of the South Brazil Bight, these areas contain rich and diverse malacofaunas that typically contain large species radiations of the tropically derived families Ovulidae, Volutidae, and Conidae. Some of the Paulinian gastropods that are endemic to the South Brazilian Bight area include the following (several illustrated in Figures 11.2, 11.3, 11.4, and 11.5):

Ovulidae
   Cyphoma guerrinii
   Cyphoma seccato
   Pseudocyphoma kathiewayae
Ranellidae
   Cabestana felipponei (also in the Uruguayan Provinciatone)
   Charonia lampas pustulata (also found in deep water in the Bahian Subprovince)
Muricidae
   Coronium elegans
   Rugotyphis cleryi
   Typhina riosi

Figure 11.4 Cone shells of the Paulinian Subprovince and Cabo Frio Region: (A) *Lamniconus clenchi* (Martins, 1943), length 43 mm. (B) *Lamniconus lemniscatus* (Reeve, 1849), length 52 mm. (C) *Lamniconus tostesi* (Petuch, 1986), length 34 mm. (D) *Lamniconus clerii* (Reeve, 1844), length 51 mm. (E, F) *Lamniconus xanthocinctus* (Petuch, 1986), length 42 mm.

Fasciolariidae
    Pleuroploca granulilabris
Volutidae
    Enaeta new species (*guildingi* of authors)
    Odontocymbiola americana
    Odontocymbiola macaensis (also found along the Cabo Frio region)
    Odontocymbiola simulatrix
Olividae
    Agaronia travassosi
    Oliva (Americoliva) circinata tostesi
    Olivancillaria urceus brasiliana
    Olivancillaria vesica vesica
Olivellidae
    Olivella defiorei
Marginellidae
    Bullata analuciae
Cancellariidae
    Cancellaria petuchi (also found off the southern Bahia coast)
Conidae
    Conasprelloides capricorni (also in the Uruguayan Provinciatone)
    Lamniconus clerii

Figure 11.5 The endemic southern Brazilian fasciolariid *Pleuroploca granulilabris* Vermeij and Snyder, 2003, a newly discovered inhabitant of the South Brazil Bight. This 57 mm specimen was dredged from 50 m depth off Florianopolis, Santa Catarina State. Although originally placed in the genus *Pleuroploca* by Vermeij and Snyder, this rarely seen species probably represents a new genus.

> *Lamniconus lemniscatus*
> *Lamniconus tostesi*
> Terebridae
> > *Strioterebrum reticulata*
> > *Strioterebrum riosi*

The presence of such a large endemic fauna of volutids and conids demonstrates that the South Brazil Bight area has retained stable, warm-temperate water conditions since at least the Pliocene. Considering the high endemicity found within the bight and the large number of geographically restricted novel taxa, the malacofauna of this large embayment most probably represents a living representation of the molluscan assemblages of the Pliocene Camachoan Paleoprovince (Petuch, 2004).

Uruguayan Provinciatone

Extending from approximately the city of Rio Grande, Rio Grande do Sul State, Brazil, south to Mar del Plata, Argentina, the Uruguayan Provinciatone also encompasses the Uruguayan and Argentinian coasts of the Rio de la Plata Estuary. The coastline of this broad area of cooler water temperatures (less than 20 degrees Celsius in the winter) is dominated by quartz sand beaches with high wave energy. Because of these low water temperatures, the dominant species radiations of the Uruguayan Provinciatone are all made up of cold-water groups. As is typical of all high-latitude provinciatonal areas, these cold-water groups occur together with a few ecologically hardy members of tropical genera, in this case *Ancilla* (Olividae), *Lamniconus* (Conidae), and *Strioterebrum* (Terebridae). Many of the gastropods that occur within the provinciatonal boundaries are very widespread taxa along the Southwestern Atlantic Shelf, often ranging from Cabo Frio, Rio de Janeiro State, Brazil, all the way south to the Golfo San Matias, Argentina, or, in a few cases, the Falkland Islands. Based on Pleistocene marine fossil faunas found on the Uruguay–Brazil border, Brazilian malacologists Lopes and Simone (2012) recently showed that the living provinciatonal malacofauna has remained essentially the same for the past one million years. Some of these extremely wide-ranging components of the Uruguayan provinciatone include the following (several illustrated in Figures 11.6, 11.7, and 11.8):

Fissurellidae
    Fissurellidea megatrema
Cassidae
    Xenophalium labiatum iheringi
Muricidae
    Coronium acanthodes
    Coronium coronatum
    Hanetia cala
    Hanetia haneti
    Muricopsis necocheana
    Trophon orbignyi
    Trophon pelseneeri
    Trophon plicatus
Buccinidae
    Buccinanops deformis
    Buccinanops duartei
    Buccinanops gradatum
    Buccinanops lamarcki
    Buccinanops uruguayensis
    Dorsanum moniliferum
Volutidae
    Adelomelon ancilla
    Adelomelon becki
    Adelomelon brasiliana
    Adelomelon (Weaveria) riosi
    Minicymbiola corderoi
    Zidona dufresnei
Olividae
    Ancilla dimidiata
    Olivancillaria carcellesi
    Olivancillaria deshayesiana
    Olivancillaria teaguei
    Olivancillaria uretai

Figure 11.6 Cone shells of the Paulinian Subprovince and Cabo Frio region: (A, B) *Sandericonus carioca* (Petuch, 1986), length 38 mm. (C, D) *Sandericonus ednae* Petuch, new species, holotype, length 29 mm (see Appendix 2). (E) *Conasprelloides* unnamed species, length 75 mm. (F) *Jaspidiconus* unnamed species, length 48 mm (*mindanus* species complex).

    Olivancillaria urceus
    Olivancillaria vesica auricularia
Olivellidae
    Olivina orejasmirandai
    Olivina plata
    Olivina puelcha
    Olivina tehuelcha
Terebridae
    Strioterebrum doellojuradoi
    Strioterebrum gemmulatum

Of particular interest in the Uruguayan Provinciatone is a large resident species radiation of the olivid genus *Olivancillaria*, with at least nine species and subspecies (Figures 11.9 and 11.10). Most of these odd olive shells live in the surf zone on open sandy beaches, where they are the major predators on sand-burrowing crustaceans such as the mole crab, *Emerita*. The surf zone along the sandy beaches of the provinciatonal area also houses a large radiation of at least five species of the buccinid genus *Buccinanops* (see Figure 11.8).

Figure 11.7 Gastropods of the Uruguayan Provinciatone: (A) *Trophon pelseneeri* Smith, 1915, length 22 mm. (B) *Trophon orbignyi* Carcelles, 1946, length 20 mm. (C) *Conasprelloides capricorni* (Van Mol, Tursch, and Kempf, 1967), length 56 mm. (D) *Lamniconus carcellesi* (Martins, 1945), length 47 mm. (E) *Xenophalium labiatum iheringi* (Carcelles, 1953), length 53 mm. (F) *Ancilla dimidiata* (Sowerby, 1850), length 23 mm.

Here, several species, in particular *Buccinanops gradatum*, are the major predators on the surf-dwelling bivalve genus *Donax*. This high-energy surf environment, with its large species radiations of molluscivorous olivids, olivellids, and buccinids, is unique within the western Atlantic.

As in all provinciatones, a number of provinciatonal endemics occur along with the tropical and nontropical faunal components. These taxa, which do not occur outside the boundaries of the transition zone, demonstrate that this area is a true provinciatone, with tropical, nontropical, and provinciatonal endemic species all occurring together in the same molluscan assemblages. Some of the Uruguayan Provinciatone endemics include the following (with one illustrated in Figures 11.6 and 11.7):

Volutidae
　　Adelomelon barratinii
　　Nanomelon vossi (off Rio Grande do Sul)
Olividae
　　Olivancillaria contortiplicata

Figure 11.8 Gastropods of the Uruguayan Provinciatone: (A) *Buccinanops gradatum* (Deshayes, 1844), length 52 mm. (B) *Buccinanops deformis* (King, 1931), length 31 mm. (C) *Fissurellidea megatrema* d'Orbigny, 1841, length 23 mm. (D) *Zidona dufresnei* (Donovan, 1823) form *affinis* Lahille, 1895, length 82 mm. (E) *Striolerebrum gemmulatum* (Kiener, 1839), length 46 mm. (F) *Dorsanum moniliferum* (Valenciennes, 1834), length 31 mm.

Olivellidae
  Orbignytesta formicacorsii
Conidae
  Lamniconus carcellesi (also found in the extreme southern part of the South Brazil Bight)

South of Mar del Plata, Argentina, the Falklands Current completely dominates the coastal waters, and the oceanographic conditions become subantarctic in nature. This shift in permanent water temperatures marks the beginning of the Patagonian Province and the end of the Uruguayan Provinciatone.

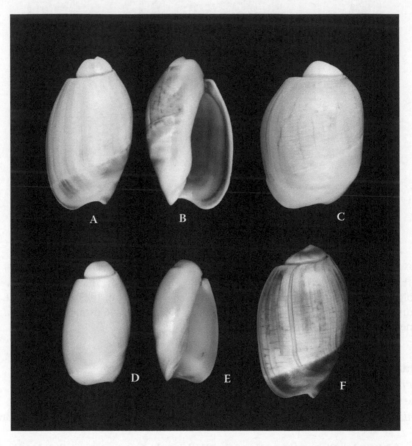

Figure 11.9 Olivid gastropods of the Uruguayan Provinciatone: (A, B) *Olivancillaria vesica vesica* (Gmelin, 1791), length 38 mm. (C) *Olivancillaria vesica auricularia* (Lamarck, 1810), length 40 mm. (D, E) *Olivancillaria teaguei* Klappenbach, 1964, length 18 mm. (F) *Olivancillaria carcellesi* Klappenbach, 1965, length 28 mm.

Figure 11.10 Olivid gastropods of the Uruguayan Provinciatone: (A, B) *Olivancillaria contortiplicata* (Reeve, 1850), length 32 mm. (C) *Olivancillaria urceus* (Röding, 1798), length 58 mm. (D) *Olivancillaria uretai* Klappenbach, 1965, length 31 mm. (E, F) *Olivancillaria deshayesiana* (Duclos, 1857), length 30 mm.

Appendix 1: Provincial Index Taxa

The 421 named species and subspecies and 21 unnamed species listed here, representing 10 separate families and subfamilies of gastropods, were used for the provincial and sub-provincial analyses shown throughout this book. These higher, family-level taxa were chosen because of their wide-ranging, amphiprovincial distributions and also because of their relatively well-established and stable systematics. As more field research is conducted in the western Atlantic and more new endemic taxa are discovered, the percentages of endemicity will undoubtedly become even higher than they are now. In the following lists, the resident subprovince of each species and subspecies is designated by a code number.

   Carolinian Province
      Georgian Subprovince = C1
      Floridian Subprovince = C2
      Suwannean Subprovince = C3
      Texan Subprovince = C4
      Yucatanean Subprovince = C5
   Caribbean Province
      Bermudan Subprovince = CR1
      Bahamian Subprovince – CR2
      Antillean Subprovince = CR3
      Nicaraguan Subprovince = CR4
      Venezuelan Subprovince = CR5
      Grenadian Subprovince = CR6
      Surinamian Subprovince = CR7
   Brazilian Province
      Cearaian Subprovince = B1
      Bahian Subprovince = B2
      Paulinian Subprovince = B3

Family Modulidae

Genus *Modulus* Gray, 1842
  Modulus bayeri B1, B2
  Modulus calusa C2
  Modulus carchedonicus CR3, CR4, CR5, CR6
  Modulus floridanus C3
  Modulus hennequini CR4 (Roatan Island; see Appendix 2)
  Modulus honkerorum B2 (see Appendix 2)
  Modulus kaicherae C1
  Modulus lindae C1
  Modulus modulus CR3, CR4, CR5, CR6
  Modulus pacei C1 (Palm Beach Provinciatone)
  Modulus new species a. C5 (similar to *calusa*)
  Modulus new species b. CR1 (endemic to Bermuda)

Family Turbinellidae

Genus *Globivasum* Abbott, 1950
  Globivasum globulus CR6 (endemic to Antigua and Barbuda)
  Globivasum globulus whicheri CR6 (endemic to Anguilla; see Appendix 2)
Genus *Turbinella* Lamarck, 1799
  Turbinella angulata C5, CR2, CR3, CR4, CR5, CR6, CR7
  Turbinella laevigata B1, B2
  Turbinella laevigata rianae CR7 (endemic to Suriname and French Guiana)
  Turbinella wheeleri C5 (named as a late Pleistocene fossil from Florida but still extant;
    listed by E. Vokes as *Turbinella scolymoides*)
Genus *Vasum* Röding, 1798
  Vasum capitellum CR3, CR6
  Vasum cassiforme B1, B2
  Vasum muricatum C2, C5, CR2, CR3, CR4, CR5, CR6, CR7

Family Conidae

(I here follow, in a modified form, the classification scheme proposed by Tucker and Tenorio, 2009)

Genus *Arubaconus* Petuch, new genus (see Appendix 2)
  Arubaconus hieroglyphus CR6 (endemic to Aruba)
Genus *Atlanticonus* Petuch and Sargent, 2011
  Atlanticonus cuna CR4 (endemic to the San Blas Islands)
  Atlanticonus glenni CR4 (endemic to the San Blas Islands)
  Atlanticonus granulatus C2, C5, CR2, CR3, CR4, CR5, CR6
  Atlanticonus ritae CR (endemic to Rosalind Bank)
Genus *Attenuiconus* Petuch, new genus (see Appendix 2)
  Attenuiconus attenuatus C2, CR2, CR3, CR4, CR5, CR6
  Attenuiconus eversoni CR4
  Attenuiconus honkeri CR5

Attenuiconus ignotus CR4
Attenuiconus poulosi CR5
Genus *Bermudaconus* Petuch, new genus (see Appendix 2)
    Bermudaconus lightbourni CR1
Genus *Brasiliconus* Petuch, new genus (see Appendix 2)
    Brasiliconus scopolorum B1
Genus *Cariboconus* Petuch, 2004
    Cariboconus bessei CR4
    Cariboconus brunneofilaris CR4
    Cariboconus deynzerorum CR3 (endemic to Banco Chinchorro Atoll)
    Caricoconus flammeacolor CR4
    Cariboconus kalafuti CR4 (endemic to Roatan Island)
    Cariboconus kirkandersi CR3 (endemic to Isla Cozumel and Islas Mujeres)
    Cariboconus magnottei CR4 (endemic to Roatan Island)
    Cariboconus sahlbergi CR2 (endemic to Bimini)
Genus *Chelyconus* Mörch, 1852
    Chelyconus ermineus C1, C2, C3, C4, C5, CR2, CR3, CR4, CR5, CR6, CR7, B1, B2, B3
Genus *Conasprelloides* Tucker and Tenorio, 2009
    Conasprelloides bajanensis CR7
    Conasprelloides brunneobandatus CR5, CR7, B1
    Conasprelloides cancellatus C2, C3, C4, C5
    Conasprelloides capricorni B3
    Conasprelloides finkli CR5
    Conasprelloides guyanensis CR7
    Conasprelloides kevani CR5
    Conasprelloides leekremeri CR2
    Conasprelloides penchaszadehi CR5
    Conasprelloides stimpsoni C1, C2, C3, C4, C5
    Conasprelloides tristensis CR5
    Conasprelloides velaensis CR5
    Conasprelloides venezuelanus CR5
    Conasprelloides villepini C1, C2, C3, C4, C5, CR1, CR2, CR3, CR4, CR5, CR6, CR7, B1, B2
    Conasprelloides new species B3
Genus *Dauciconus* Cotton, 1945
    Dauciconus amphiurgus C1, C2, C3, C4, C5
    Dauciconus aureonimbosus C3
    Dauciconus boui CR6 (endemic to Martinique)
    Dauciconus daucus C2, C5, CR2, CR3, CR4, CR5, CR6
    Dauciconus glicksteini C1 (Palm Beach Provinciatone)
    Dauciconus goajira CR5
    Dauciconus jorioi B2 (endemic to Trindade Island; see Appendix 2)
    Dauciconus norai CR6 (endemic to Martinique)
    Dauciconus riosi B2
    Dauciconus vikingorum CR5
    Dauciconus worki B2
Genus *Gladioconus* Tucker and Tenorio, 2009
    Gladioconus mus C1, C2, C4, C5, CR1, CR2, CR3, CR4, CR5, CR6, CR7

Genus *Gradiconus* da Motta, 1991
    Gradiconus anabathrum C3
    Gradiconus anabathrum tranthami (= *antoni, tortuganus*) C2
    Gradiconus aureopunctatus CR4
    Gradiconus bayeri CR4 (endemic to the San Blas Islands)
    Gradiconus burryae C2
    Gradiconus cingulatus CR5
    Gradiconus ernesti CR4
    Gradiconus garciai CR4
    Gradiconus gibsonsmithorum CR5
    Gradiconus largillierti C4
    Gradiconus maya C5
    Gradiconus mazzolii C2
    Gradiconus ostrinus CR4
    Gradiconus paraguana CR5
    Gradiconus parascalaris CR5
    Gradiconus paschalli CR4
    Gradiconus paulae CR5
    Gradiconus philippii C1
    Gradiconus portobeloensis CR4
    Gradiconus rosemaryae CR4
    Gradiconus sennottorum C5
Genus *Kellyconus* Petuch, new genus (see Appendix 2)
    Kellyconus binghamae C1 (Palm Beach Provinciatone)
    Kellyconus patae C2, CR2, CR3, CR4, CR5, CR6
    Kellyconus rachelae CR7
Genus *Lamniconus* da Motta, 1991
    Lamniconus carcellesi B3
    Lamniconus clenchi B3
    Lamniconus clerii B3
    Lamniconus lemniscatus B3
    Lamniconus tostesi B3
    Lamniconus xanthocinctus B3
Genus *Lindaconus* Petuch, 2002
    Lindaconus atlanticus C1, C2, C3, C4
    Lindaconus aureofasciatus C3
    Lindaconus baylei (= *spurius arubaensis*) CR5
    Lindaconus lindae CR2 (endemic to deep-water western Bahamas)
    Lindaconus lorenzianus CR4
    Lindaconus phlogopus CR5
    Lindaconus spurius CR2, CR3, CR4, CR5, CR6
    Lindaconus sunderlandi CR4
    Lindaconus therriaulti C5 (see Appendix 2)
Genus *Magelliconus* da Motta, 1991
    Magelliconus cidaris CR3
    Magelliconus eleutheraensis CR2 (see Appendix 2)
    Magelliconus explorator CR3
    Magelliconus exquisitus CR3
    Magelliconus jacarusoi CR2

Magelliconus maculiferus CR6
Magelliconus magellanicus CR3, CR6
Magelliconus pseudocardinalis B2
Magelliconus sphaecelatus CR3
Magelliconus zylmanae CR2
Genus *Poremskiconus* Petuch, new genus (see Appendix 2)
Poremskiconus abrolhosensis (= *baiano*) B2
Poremskiconus archetypus B2
Poremskiconus beddomei CR6 (Grenadines)
Poremskiconus bertarollae B2
Poremskiconus brasiliensis B2
Poremskiconus cargilei B2
Poremskiconus colombi CR6 (endemic to Martinique)
Poremskiconus colombianus CR5 (Rosarios Islands, Colombia)
Poremskiconus edwardpauli CR4 (endemic to San Blas Islands)
Poremskiconus hennequini CR6 (endemic to Martinique)
Poremskiconus mauricioi B1
Genus *Purpuriconus* da Motta, 1991
Purpuriconus abbotti CR2 (endemic to Eleuthera Island)
Purpuriconus alainalaryi CR5 (Islas Rosarios and islands off the Golfo de Morrosquillo)
Purpuriconus arangoi CR2 (Cuba and the Bahamas)
Purpuriconus belizeanus CR3 (endemic to Glover's Atoll, Belize)
Purpuriconus cardinalis CR3
Purpuriconus caysalensis CR2 (endemic to Cay Sal Bank)
Purpuriconus dianthus CR2 (Turks and Caicos)
Purpuriconus donnae CR2 (endemic to the Bimini Chain)
Purpuriconus harasewychi CR2 (endemic to Sale Cays)
Purpuriconus havanensis CR2 (endemic to northern Cuba)
Purpuriconus hilli CR4 (endemic to the San Blas Islands)
Purpuriconus inconstans CR3 (= *mayaguensis*)
Purpuriconus jucundus CR2 (endemic to the Abacos)
Purpuriconus kulkulcan CR4
Purpuriconus liratus CR2 (Turks and Caicos)
Purpuriconus lucaya CR2 (endemic to western Grand Bahama Island)
Purpuriconus olgae CR2 (endemic to northwestern Cuba)
Purpuriconus ortneri CR2 (Exuma Sound)
Purpuriconus primula CR2 (Turks and Caicos)
Purpuriconus richardbinghami CR2 (Bimini Chain)
Purpuriconus rosalindensis CR4
Purpuriconus speciosissimus CR3
Purpuriconus stanfieldi CR2
Purpuriconus theodorei CR2
Purpuriconus new species 1 CR3 (cf. *kulkulcan*; endemic to Samana, Dominican Republic)
Purpuriconus new species 2 CR3 (cf. *cardinalis*; endemic to Jamaica)
Genus *Sandericonus* Petuch, new genus (see Appendix 2)
Sandericonus carioca B2, B3
Sandericonus ednae B3 (Cabo Frio Region; see Appendix 2)
Sandericonus hunti CR6
Sandericonus knudseni CR6

      Sandericonus perprotractus CR7
      Sandericonus sanderi CR6
      Sandericonus sorenseni CR6
Genus *Stephanoconus* Mörch, 1852
      Stephanoconus regius C1, C2, C4, C5, CR1, CR2, CR3, CR4, CR5, CR6, CR7, B1, B2, B3
Genus *Tenorioconus* Petuch and Drolshagen, 2011
      Tenorioconus aurantius CR6 (endemic to Curacao and Bonaire)
      Tenorioconus caracanus CR5 (Isla Margarita area)
      Tenorioconus cedonulli CR6 (endemic to St. Vincent)
      Tenorioconus curassaviensis CR6 (endemic to Aruba)
      Tenorioconus dominicanus CR6 (endemic to the Grenadines)
      Tenorioconus duffyi CR6 (endemic to Los Roques Atoll)
      Tenorioconus granarius CR5
      Tenorioconus grenadensis CR6 (endemic to Grenada and the Grenadines)
      Tenorioconus harlandi CR4
      Tenorioconus insularis CR6 (endemic to St. Lucia)
      Tenorioconus jesusramirezi CR5
      Tenorioconus juliandrieae CR4
      Tenorioconus mappa (Venezuela, Trinidad and Tobago, Barbados)
      Tenorioconus mappa sanguineus CR5 (endemic to the Goajira Peninsula)
      Tenorioconus martinicanus CR6 (endemic to Martinique)
      Tenorioconus panamicus CR4 (endemic to the San Blas Islands)
      Tenorioconus pseudoaurantius CR6 (endemic to Grenada and the Grenadines)
      Tenorioconus trinitarius CR6 (Isla Margarita)
Genus *Tuckericonus* Petuch, new genus (see Appendix 2)
      Tuckericonus bahamensis CR2
      Tuckericonus cerrutii C4
      Tuckericonus flamingo C1 (Palm Beach Provinciatone)
      Tuckericonus flavescens C1, C2, CR2
      Tuckericonus new species (similar to *flavescens*, but with multi-whorl protoconch) C1

Family Conilithidae

(I here follow, in a modified form, the classification scheme proposed by Tucker and Tenorio, 2009)

Genus *Artemidiconus* da Motta, 1991
      Artemidiconus selenae B1
      Artemidiconus yemanjae B2
Genus *Coltroconus* Petuch, new genus (see Appendix 2)
      Coltroconus delucai B2
      Coltroconus iansa (= *bodarti*) B2
      Coltroconus schirrmeisteri B2
Genus *Dalliconus* Tucker and Tenorio, 2009
      Dalliconus armiger C4
      Dalliconus coletteae CR6
      Dalliconus kremerorum CR6 (endemic to Barbados)
      Dalliconus lenhilli CR3
      Dalliconus macgintyi C1, C2, C3, C4, C5

Dalliconus mazei CR6

Dalliconus pacei CR2

Dalliconus rainseae C5

Dalliconus roberti CR3

Dalliconus sauros C4

Dalliconus new species (*"macgintyi"* from Brazil) B2, B3, B4

Genus *Jaspidiconus* Petuch, 2004

Jaspidiconus acutimarginatus C2

Jaspidiconus agassizii CR3 (deep water)

Jaspidiconus allamandi CR4 (endemic to Roatan Island; see Appendix 2)

Jaspidiconus anaglypticus CR3 (endemic to Puerto Rico)

Jaspidiconus branhamae CR2 (endemic to the Abacos)

Jaspidiconus damasoi B1

Jaspidiconus duvali CR3, CR6

Jaspidiconus exumaensis C2 (Exuma Sound; see Appendix 2)

Jaspidiconus fluviamaris C1 (Palm Beach Provinciatone), C2

Jaspidiconus henckesi B2

Jaspidiconus jaspideus CR5, CR6

Jaspidiconus mackintoshi CR3 (endemic to the Virgin Islands; see Appendix 2)

Jaspidiconus mindanus C1, C2, C3, C4, C5, CR2, CR3, CR4, CR5, CR6, CR7, B1, B2, B3

Jaspidiconus mindanus bermudensis CR1 (endemic to Bermuda)

Jaspidiconus nodiferus CR2

Jaspidiconus oleiniki B2 (endemic to the Bimini Chain; see Appendix 2)

Jaspidiconus pealii C2

Jaspidiconus pfluegeri C1, also Palm Beach Provinciatone

Jaspidiconus pusillus B1, B2, B3

Jaspidiconus roatanensis CR4 (endemic to Roatan Island)

Jaspidiconus sargenti CR4 (see Appendix 2)

Jaspidiconus stearnsi C3

Jaspidiconus vanhyningi C1 (Palm Beach Provinciatone), C2, C3

Jaspidiconus verrucosus CR2

Jaspidiconus new species a. (*"agassizii"* from Brazil) B2, B3

Jaspidiconus new species b. (*"mindanus"* from Brazil) B2

Jaspidiconus new species c. (*"jaspideus"* from Brazil B1, B2

Jaspidiconus new species d. B2

Jaspidiconus new species e. B2

Genus *Kohniconus* Tucker and Tenorio, 2009

Kohniconus centurio CR2, CR3, CR4, CR5, CR6, CR7, B1, B2, B3

Kohniconus delessertii C1, C2, C3, C4, C5

Kohniconus janowskyae C5, CR4, CR5

Genus *Perplexiconus* Tucker and Tenorio, 2009

Perplexiconus puncticulatus (and forms *cardonensis* and *mauritianus*) CR5

Perplexiconus puncticulatus columba CR6

Family Muricidae

Subfamily Muricinae

(The western Atlantic representatives of the Subfamilies Muricopsinae, Rapaninae, Ergalitaxinae, Ocenebrinae, Typhinae, and Trophoninae are not included here.)

Genus *Aspella* Mörch, 1877
 Aspella castor C2, C5, CR2, CR3, CR4, CR5, CR6
 Aspella cryptica B2
 Aspella morchi B2
 Aspella senex C1, C2, C3, C4, C5
Genus *Attiliosa* Emerson, 1968
 Attiliosa aldridgei C2, C5, CR2, CR3, CR6
 Attiliosa bessei CR4 (Rosalind Bank)
 Attiliosa glennduffyi CR3, CR6
 Attiliosa kevani CR3, CR4
 Attiliosa perplexa B1, B2, B3
 Attiliosa philippiana C2, C5, CR4, CR5
 Attiliosa striatoides C2, C5
 Attiliosa new species B2
Genus *Calotrophon* Hertlein and Strong, 1951
 Calotrophon andrewsi C2, C3, C5
 Calotrophon hystrix C3, C4, C5
 Calotrophon ostrearum C2, C3, C5
Genus *Chicoreus* Montfort, 1810
 Chicoreus brevifrons CR3, CR4, CR5, CR6, CR7
 Chicoreus brevifrons franchii CR5
 Chicoreus bullisi CR4
 Chicoreus cosmani CR3
 Chicoreus dilectus C1, C2, C3, C4, C5
 Chicoreus dunni CR2
 Chicoreus emilyae CR4
 Chicoreus florifer CR2, CR4
 Chicoreus hilli CR4 (San Blas Islands)
 Chicoreus mergus C2, C5, CR2, CR3, CR4, CR5, CR6, CR7, B1
 Chicoreus rachelcarsonae C3
 Chicoreus spectrum CR6, CR7, B1, B2
Genus *Dermomurex* Monterosato, 1890
 Dermomurex alabastrum CR3, CR6
 Dermomurex binghamae CR2
 Dermomurex columbi CR6
 Dermomurex coonsorum CR3 (endemic to Glover's Atoll; see Appendix 2)
 Dermomurex elizabethae C2, CR1, CR2, CR3
 Dermomurex glicksteini C1 (Palm Beach Provinciatone)
 Dermomurex kaicherae CR5, CR6
 Dermomurex morchi B2
 Dermomurex olssoni CR2, CR3
 Dermomurex pacei C2

Dermomurex pauperculus C1, C2, C3, C4, C5, CR1, CR2, CR3, CR4, CR5, CR6
Dermomurex sarasuae CR2 (Cuba)
Dermomurex worsfoldi CR2
Dermomurex new species a. CR1 (endemic to Bermuda)
Subgenus *Trialatella* Berry, 1964
    Dermomurex abyssicola CR3, CR4, CR5, CR6
    Dermomurex cuna CR4 (= *"antecessor"*) (San Blas Islands)
    Dermomurex leali B2
    Dermomurex oxum B2
Genus *Hexaplex* Perry, 1810
    Hexaplex strausi CR6 (Dominica, Martinique)
Genus *Panamurex* Woodring, 1959
    Panamurex carnicolor CR6
    Panamurex eugeniae CR5
    Panamurex harasewychi CR5 (originally described as a *"Murexsul"*)
    Panamurex petuchi CR5
    Panamurex velero CR5
Genus *Pazinotus* E. Vokes, 1970
    Pazinotus bodarti B2
    Pazinotus goesi CR3
    Pazinotus stimpsoni C1, C2, C3, C4, C5, C6, CR2, CR3, CR4, CR5, CR6
Genus *Pyllonotus* Swainson, 1833
    Phyllonotus margaritensis CR5
    Phyllonotus mexicanus C5
    Phyllonotus oculatus C2, C5, CR2, CR3, CR4, CR5, CR6, CR7, B1, B2, B3
    Phyllonotus pomum C1, C2, C3, C4, C5, CR1, CR2, CR3, CR4, CR5, CR6, CR7
    Phyllonotus pomum whymani (may be a full species) C3
    Phyllonotus new species CR5 (similar to *P. marguritensis*)
Genus *Poirieria* Jousseaume. 1880
    Poirieria actinophorus C5, CR2, CR3, CR4, CR5, CR6, CR7, B1, B2, B3
    Poirieria atlantis CR2 (Cuba)
    Poirieria hystricina CR2, CR3, CR6
    Poirieria nuttingi C3
    Poirieria oregonia B1
    Poirieria pazi C1, C2, C3, C4, C5, CR2, CR4
Genus *Pterochelus* Jousseaume, 1880
    Pterochelus ariomus C2
Genus *Pteropurpura* Jousseaume, 1880
    Pteropurpura bequaerti C1, C2, C3, C4, C5
Genus *Pterynotus* Swainson, 1833
    Pterynotus bushae C2, CR2
    Pterynotus emilyae CR2 (Cuba)
    Pterynotus guesti C2 (endemic to the Florida Keys)
    Pterynotus havanensis CR2
    Pterynotus phaneus C1, C2
    Pterynotus phyllopterus CR6
    Pterynotus radwini CR3 (endemic to Belize)
    Pterynotus tristichus CR2, CR3, CR4, CR5, CR6, CR7, B1, B2
    Pterynotus xenos CR3, CR4

Genus *Siratus* Jousseaume. 1880
   Siratus aguayoi CR2 (Cuba)
   Siratus articulatus CR2, CR3, CR4, CR6
   Siratus beaui C1, C2, C3, C4, C5, CR1, CR2, CR3, CR4, CR5, CR6, CR7, B1, B2, B3
   Siratus cailleti CR2, CR3, CR6
   Siratus caudacurta C4, C5
   Siratus ciboney CR2, CR3, CR6
   Siratus coltrorum B2, B3
   Siratus colellai CR3
   Siratus finlayi CR2 (Cuba)
   Siratus formosus CR2, CR3, CR4, CR5, CR6, CR7
   Siratus guionetti CR6
   Siratus hennequini CR4
   Siratus kugleri CR3, CR6
   Siratus lamyi CR7
   Siratus motacilla CR6
   Siratus perelegans CR6
   Siratus senegalensis B1, B2, B3
   Siratus springeri CR7, B1
   Siratus tenuivaricosus B1, B2, B3
   Siratus thompsoni CR7, B1
   Siratus vokesorum CR2
   Siratus yumurinus CR2 (Cuba and Bahamas)
Genus *Timbellus* de Gregorio, 1885
   Timbellus lightbourni CR1 (endemic to Bermuda)
Genus *Vokesimurex* Petuch, 1994
   Vokesimurex anniae C5
   Vokesimurex bayeri CR5
   Vokesimurex bellus CR3, CR5
   Vokesimurex blakeanus CR5, CR6
   Vokesimurex cabritti C1, C2, C3, C4, C5, CR2, CR3
   Vokesimurex carolinae B2
   Vokesimurex chrysostomus CR6
   Vokesimurex donmoorei CR5
   Vokesimurex garciai CR4
   Vokesimurex hammani CR6
   Vokesimurex lindajoyceae C3
   Vokesimurex messorius CR5, CR6, CR7
   Vokesimurex morrisoni (= "bellegladeensis") C1
   Vokesimurex rubidus C1, C2, C3, C4, C5
   Vokesimurex rubidus marcoensis C2 (endemic to Florida Bay)
   Vokesimurex rubidus panamicus CR4
   Vokesimurex sallasi C5
   Vokesimurex samui CR4
   Vokesimurex sunderlandi CR5
   Vokesimurex woodringi CR5

Family Fasciolariidae

Subfamily Fasciolariinae

(The western Atlantic representatives of the subfamilies Peristerniinae and Fusininae are not included here.)

Genus *Cinctura* Hollister, 1957
Cinctura branhamae C5
Cinctura hunteria C1, C2, C3, C4, C5
Cinctura hunteria keatonorum C1 (see Appendix 2)
Cinctura lilium C4
Cinctura totuganum C3
Genus *Fasciolaria* Lamarck, 1799
Fasciolaria agatha B2
Fasciolaria bullisi C5
Fasciolaria tephrina C4, C5
Fasciolaria tulipa C1, C2, C3, C4, C5, CR1, CR2, CR3, CR4, CR5, CR6, CR7, B1
Fasciolaria tulipa hollisteri C5
Genus *Pleuroploca* Fischer, 1884
Pleuroploca aurantiaca B1, B2
Pleuroploca granulilabris B3
Genus *Triplofusus* Olsson and Harbison, 1953
Triplofusus papillosus (= *giganteus*) C1, C2, C3, C4, C5

Family Volutidae

Subfamily Lyriinae

(The western Atlantic representatives of the subfamilies Volutinae, Scaphellinae, Zidoniinae, Odontocymbiolinae, and Plicolivinae are not included here.)

Genus *Enaeta* H. and A. Adams, 1853
Enaeta bessei CR4 (endemic to Rosalind Bank; see Appendix 2)
Enaeta cylleniformis CR2
Enaeta guildingi CR6
Enaeta leonardhilli B1
Enaeta lindae CR2 (endemic to the Bimini Chain; see Appendix 2)
Enaeta reevei CR4
Enaeta new species a. C2 (cf. *cylleniformis* from Biscayne Bay and northern Florida Keys)
Enaeta new species b. B1, B2 (endemic to coastal Brazil)
Enaeta new species c. B3 ("guildingi" of authors)
Genus *Lyria* Gray, 1847
Lyria archeri CR6
Lyria beaui CR6
Lyria cordis CR3
Lyria leonardi CR3, CR5, CR6
Lyria russjenseni CR3
Lyria sabaensis CR3
Lyria vegai CR3

Family Olividae

Subfamily Olivinae

(The western Atlantic representatives of the subfamilies Ancillinae and Agaroniinae are not included here.)

Genus *Oliva* Bruguiere, 1789
Subgenus *Americoliva* Petuch, new subgenus (see Appendix 2)
Oliva antillensis CR3, CR4, CR6
Oliva bahamasensis CR2 (endemic to Bimini)
Oliva barbadensis CR6 (endemic to Barbados)
Oliva bewleyi CR5
Oliva bifasciata C2, C3, C4, C5, CR2, CR3, CR4, CR5, CR6
Oliva bifasciata bollingi
Oliva bifasciata jenseni CR1 (endemic to Bermuda)
Oliva broderipi CR3 (Hispaniola)
Oliva broderipi subspecies CR3 (Puerto Rico)
Oliva circinata B2
Oliva circinata jorioi B1 (see Appendix 2)
Oliva circinata tostesi B3
Oliva contoyensis C5
Oliva drangai CR6 (endemic to Trinidad and Tobago)
Oliva figura CR6
Oliva fugurator CR5 (Venezuela)
Oliva fulgurator bullata CR6
Oliva fulgurator fusiforme CR6 (Aruba)
Oliva goajira CR5
Oliva jamaicensis CR3 (endemic to Jamaica)
Oliva maya C5
Oliva mooreana CR3 (deep water off Belize; see Appendix 2)
Oliva obesina CR5
Oliva oblonga CR5
Oliva olivacea CR6
Oliva porcea CR5
Oliva reclusa CR
Oliva reticularis CR2, CR3, CR6
Oliva reticularis ernesti CR4
Oliva sargenti CR6 (endemic to Aruba)
Oliva sayana C1
Oliva sayana sarasotaensis C3
Oliva sayana texana C4
Oliva sunderlandi C3
Oliva new species CR7 (similar to *circinata*)
Subgenus *Cariboliva* Petuch and Sargent, 1986
Oliva scripta CR2, CR3, CR4, CR5, CR6, CR7, B1
Oliva scripta venezuelana CR5

(**Note:** Recent research by Pierre Recourt, The Netherlands, has shown that there are at least 10 more unnamed species of Caribbean *Americola*.)

Family Cancellariidae
Subfamily Cancellariinae

(The western Atlantic representatives of the subfamily Admetinae are not included here.)

Genus *Aphera* H. and A. Adams, 1854
   Aphera lindae CR6 (endemic to Barbados)
Genus *Axelella* Petit, 1988
   Axelella agassizii C1, C2, C3, C4, C5
   Axelella brasiliensis B2, B3
   Axelella epomis CR5
   Axelella smithi C1, C2, C3, C4, C5, CR2, CR4, CR5
Genus *Bivetopsia* Jousseaume, 1887
   Bivetopsia rugosum C2, C5, CR2, CR3, CR4, CR5, CR6
Genus *Cancellaria* Lamarck, 1799
   Cancellaria adelae C2
   Cancellaria mediamericana CR4
   Cancellaria petuchi B2, B3
   Cancellaria reticulata C1, C2, C3, C4, C5, CR2, CR3, CR5, CR7
   Cancellaria richardpetiti C3
   Cancellaria rosewateri C4 (deep-water petroleum seeps)
Genus *Gerdiella* Olsson and Bayer, 1973
   Gerdiella alvesi B1, B2
   Gerdiella cingulata C2, C3, C4, C5, CR2, CR3
   Gerdiella gerda CR2, CR3, CR4, CR5, CR6
   Gerdiella santa C1, C2, C3, C4, C5
Genus *Gergovia* Cossman, 1899
   Gergovia petiti B2
Genus *Trigonostoma* Blainville, 1827
   Trigonostoma tenerum C1, C2, C3, C4, C5

Family Cancellariidae
Subfamily Plesiotritoninae

Genus *Tritonoharpa* Dall, 1908
   Tritonoharpa bayeri CR5
   Tritonoharpa cubapatriae CR2 (endemic to northern Cuba)
   Tritonoharpa janowskyi C1 (Palm Beach Provinciatone)
   Tritonoharpa lanceolata C1, C2, C3, C4, C5, CR1, CR2, CR3, CR4, CR5, CR6, CR7
   Tritonoharpa leali B1, B2, B3
   Tritonoharpa new species CR4 (endemic to the San Blas Islands)

Appendix 2: Additions to western Atlantic molluscan biodiversity

While compiling the taxonomic data for the provincial and subprovincial analyses used throughout this book, I found that several important new genera and species were still undescribed. As many of these belong to the families and subfamilies of the 10 key provincial index taxa that are used to define provincial boundaries, I felt that it was important for these indicator mollusks to be formally described. In total, 31 new species and subspecies and 11 new genera and subgenera are described here. The holotypes of the new species and subspecies are deposited either in the type collection in the Department of Malacology at the Field Museum of Natural History in Chicago, Illinois (and bear FMNH numbers) or in the type collection in the Department of Malacology at the Los Angeles County Museum of Natural History in California (and bear LACM numbers). These include the following.

New species and subspecies

Planaxidae
   Planaxis (Supplanaxis) nancyae new species (FMNH 328402) (Antillean Subprovince)
Modulidae
   Modulus hennequini new species (FMNH 328403) (Nicaraguan Subprovince)
   Modulus honkerorum new species (FMNH 328404) (Bahamian Subprovince)
Cypraeidae
   Macrocypraea cervus lindseyi new subspecies (LACM 3222) (Antillean Subprovince)
Triviidae
   Pusula bessei new species (FMNH 328408) (Nicaraguan Subprovince)
Muricidae
   Dermomurex coonsorum new species (FMNH 328409) (Antillean Subprovince)
   Murexiella deynzerorum new species (FMNH 328410) (Bahamian Subprovince)
   Muricopsis honkeri new species (FMNH 328411) (Bahamian Subprovince)

Roquesia lindae new species (FMNH 328412) (Grenadian Subprovince)
Stramonita buchecki new species (FMNH 328413) (Georgian Subprovince)
Fasciolariidae
Cinctura hunteria keatonorum new subspecies (FMNH 328415) (Georgian Subprovince)
Polygona bessei new species (FMNH 328416) (Nicaraguan Subprovince)
Polygona paulae new species (FMNH 328417) (Bahamian Subprovince)
Busyconidae
Fulguropsis spiratum keysensis new subspecies (FMNH 328418) (Floridian Subprovince)
Buccinidae
Engina dicksoni new species (FMNH 328420) (Yucatanean Subprovince)
Turbinellidae
Globivasum globulus whicheri new subspecies (FMNH 328423) (Grenadian Subprovince)
Volutidae
Enaeta bessei new species (FMNH 328424) (Nicaraguan Subprovince)
Enaeta lindae new species (FMNH 328425) (Bahamian Subprovince)
Olividae
Oliva (Americoliva) circinata jorioi new subspecies (FMNH 328428) (Cearaian Subprovince)
Oliva (Americoliva) mooreana new species (FMNH 328429) (Antillean Subprovince)
Conidae
Dauciconus jorioi new species (FMNH 328432) (Bahian Subprovince)
Lindaconus therriaulti new species (FMNH 328433) (Yucatanean Subprovince)
Magelliconus eleutheraensis new species (FMNH 328434) (Bahamian Subprovince)
Sandericonus ednae new species (FMNH 328431) (Paolinian Subprovince)
Conilithidae
Dalliconus coletteae new species (FMNH 328436) (Grenadian Subprovince)
Jaspidiconus allamandi new species (FMNH 328435) (Nicaraguan Subprovince)
Jaspidiconus exumaensis new species (FMNH 328437) (Bahamian Subprovince)
Jaspidiconus mackintoshi new species (FMNH 328439) (Antillean Subprovince)
Jaspidiconus oleiniki new species (FMNH 328440) (Bahamian Subprovince)
Jaspidiconus sargenti new species (FMNH 328441) (Nicaraguan Subprovince)
Bivalvia veneridae
Mercenaria hartae new species (FMNH 328438) (Palm Beach Provinciatone)

New genera

Muricidae-Ergalitaxinae
Roquesia new genus (type *Roquesia lindae*) (Grenadian Subprovince restricted)
Olividae-Olivinae
Americoliva new subgenus (type: *Oliva sayana*) (Carolinian, Caribbean, Brazilian, Panamic Provinces)
Conidae-Puncticulinae
Arubaconus new genus (type: *Conus hieroglyphus*) (Grenadian Subprovince; Aruba only)
Attenuiconus new genus (type: *Conus attenuatus*) (widespread Caribbean Province)
Bermudaconus new genus (type: *Conus lightbourni*) (Bermudan Subprovince restricted)
Brasiliconus new genus (type: *Conus scopolorum*) (Brazilian Province restricted)
Kellyconus new genus (type: *Conus patae*) (widespread Caribbean Province)
Poremskiconus new genus (type: *Conus archetypus*) (Grenadian Subprovince, Caribbean Province, and widespread Brazilian Province)

Sandericonus new genus (type: *Conus carioca*) (Grenadian Subprovince, Caribbean Province, and widespread Brazilian Province)

Tuckericonus new genus (type: *Conus flavescens*) (widespread Caribbean Province)
Conilithidae-Conilithinae

Coltroconus new genus (type: *Conus iansa*) (Bahian Subprovince, Brazilian Province)

All of these taxa are described below and are here arranged phylogenetically.

Descriptions of new species, subspecies, genera, and subgenera

Gastropoda
Prosobranchia
Cerithioidea

Family Planaxidae
Genus *Planaxis* Lamarck, 1822
Subgenus *Supplanaxis* Thiele, 1929

Planaxis (Supplanaxis) nancyae new species
(Figure 6.13A,B)

Description: Shell of average size for genus, slender, smooth, devoid of deeply impressed sulci or grooves, highly polished and glossy; anterior tip with 3–4 very faint, thin, low spiral cords; shoulder sloping; spire elevated, with slightly convex whorls and with single faint subsutural groove around early whorls only; suture impressed; aperture proportionally large, flaring, oval in shape, with 8–10 large, low, rounded teeth along edge of lip; columellar area concave, with single large callus-like tooth at posterior end; shell color uniform dark reddish-brown; aperture edged with white, turning reddish-brown farther inside.

Holotype: Length 11 mm, 7 mm, FMNH 328402; **Paratype:** length 11 mm, same locality as holotype, in the research collection of the author.

Type Locality: Collected under rocks in 2 m depth off southern Gonave Island, Haiti.

Etymology: Named for Nancy Long of North Carolina, amateur naturalist and award-winning elementary school teacher.

Discussion: The new species is the second *Planaxis (Supplanaxis)* known from the tropical western Atlantic and represents a previously overlooked taxon. *Planaxis (Supplanaxis) nancyae* is most similar to the common, widespread intertidal *P. (Supplanaxis) nucleus* (Bruguiere, 1789), which ranges across southern Florida and the West Indies (Warmke and Abbott, 1962: 70) but differs in being a smaller, more slender, and less inflated shell; in having a proportionally higher and narrower spire; in being highly polished and shiny; in lacking the strong spiral cords and deeply incised spiral grooves around the shoulder, subsutural area, and anterior end (as seen in *nucleus*); in having weaker and less developed teeth on the inner edge of the lip; and in having a reddish-brown shell color instead of black or dark brown-black.

Planaxis (Supplanaxis) nancyae lives in a completely different environment from *P. (Supplanaxis) nucleus*, being found in subtidal depths of 2–5 m, where it lives on algae-covered rocks or under coral rubble. *Planaxis (Supplanaxis) nucleus*, on the other hand, is an intertidal species, preferring open rock surfaces, often in surf conditions, where it is frequently exposed at low tide. Although the type lot was collected in Haiti, specimens similar to *P. (Supplanaxis) nancyae* have been collected on St. Johns, Virgin Islands (M. Coltro, personal communication). The new species may prove to be found in deep subtidal areas all along the Greater Antilles Arc, from Hispaniola to the Virgin Islands. The new deeper water species also appears to be less common than its widespread intertidal relative.

Family Modulidae
Genus *Modulus* Gray, 1842

Modulus hennequini new species
(Figure 7.11A)

Description: Shell of average size for genus, dorso-ventrally flattened, subturbinate, with smooth, waxy appearance; spire low and flattened, with earliest whorls projecting and sub-pyramidal; shell uniformly pale yellow-tan; shoulder carinated, sharp, ornamented with 7 large, flattened knobs that are rounded on their edges; knobs extend onto subsutural area and spire whorls as strong, sharp ribs; spire whorls ornamented with 4–5 very faint, thin spiral threads; base of shell ornamented with 4 large, coarse spiral cords; basal cords faintly ornamented with tiny beads and pustules; aperture proportionally large, with white interior and numerous coarse ribs that extend well into interior; columella with single large, sharp tooth at extreme anterior end; umbilicus small, round, distinctly perforate.

Holotype: Width 12 mm, height 7 mm, FMNH 328403.

Type Locality: In coral rubble, 20 m depth off Coxon's Hole, Roatan Island, Honduras.

Etymology: Named for Francis Hennequin, renowned diver, collector, and Caribbean naturalist, who collected the holotype off Roatan Island.

Discussion: Of the known western Atlantic *Modulus* species, the new Honduran species is most similar to *Modulus kaicherae* Petuch, 1987 from the Georgian subprovince of the Carolinian province. Although having similar large shoulder knobs and an astraeiform appearance, *Modulus hennequini* differs from its northern congener in having a more flattened, less turbinate shell, in having proportionally larger shoulder knobs, in lacking the strong sculpture of spiral cords that cover the shell surface, and in being a smoother shell with a waxy appearance.

Modulus honkerorum new species
(Figure 6.7A,B)

Description: Shell small for genus, very rounded, globose, shiny and polished, porcellaneous; spire slightly elevated, subpyramidal and stepped; shoulder at midbody of shell, angled, bordered by large rounded cord; subsutural area and shoulder smooth and shiny, ornamented with 10–12 low, rounded knobs; base of shell ornamented with 3 large rounded spiral cords; umbilicus small, deep, open, bordered by single large, thick spiral cord; parietal area at base of umbilicus with single large tooth that extends into aperture; aperture round; edge of lip faintly crenulated; shell color pure snow white, with pale tan early whorls and protoconch; parietal and columellar areas and parietal tooth colored pale lavender purple; interior of aperture pure white.

Holotype: Width 9 mm, height 8 mm, FMNH 328404; **Paratype:** width 8 mm, same locality as holotype, in the research collection of the author.

Type Locality: On coral rubble in 1 m depth, on small spoil island off Tarpum Bay, Exuma Sound, Eleuthera Island, Great Bahama Bank, Bahamas.

Etymology: Named for Thomas and Paula Honker, renowned divers and collectors, of Delray Beach, Florida, in recognition of their many discoveries of new Bahamian mollusks.

Discussion: This small, globose modulid is unlike any other known species from the western Atlantic. With its distinctive shell morphology, *Modulus honkerorum* could be confused only with the widespread Caribbean *Modulus modulus* (Linnaeus, 1758). The new Bahamian species differs from *M. modulus* in being a smaller, much more globose and rounded shell; in lacking large, sharply pointed shoulder knobs; in being pure white and lacking any dark color patches and stripes (as on *modulus*); and in having a shiny, polished, porcellaneous shell texture. At present, *Modulus honkerorum* has been found only on Eleuthera but could be present around the entire Exuma Sound and larger islands to the south.

The Exuma Sound species is the newest known member of the rich *Modulus* species radiation the tropical western Atlantic, the largest known from anywhere on Earth. To date, 11 species have been described from the Carolinian, Caribbean, and Brazilian provinces (Petuch, 1987, 1988, 2001):

Modulus bayeri Petuch, 2001; Brazil
Modulus calusa Petuch, 1988; Florida Bay, southeastern Florida, and possibly Yucatan
Modulus carchedonicus (Lamarck, 1822); Greater Antilles to South America
Modulus floridanus Conrad, 1869; western Florida
Modulus hennequini Petuch, new species; Roatan Island
Modulus honkerorum Petuch, new species; Eleuthera Is., Bahamas
Modulus kaicherae Petuch, 1987; deep water, Georgia to southeastern Florida
Modulus lindae Petuch, 1987; deep water, North Carolina to southeastern Florida
Modulus modulus (Linnaeus, 1758); Greater and Lesser Antilles and northern South America
Modulus pacei Petuch, 1988; southeastern Florida
Modulus perlatus (Gmelin, 1791); Greater Antilles

Cypraeoidea

Family Cypraeidae
Subfamily Cypraeinae
Genus *Macrocypraea* Schilder, 1930
Subgenus *Lorenzicypraea* Petuch and Drolshagen, 2011

Macrocypraea (Lorenzicypraea) cervus lindseyi new subspecies
(Figure 6.15A,B,C)

Description: Shell small for species, heavy and thickened, inflated, subcylindrical; posterior processes well developed, thickened, projecting; auricles proportionally large, distinctly flanged; base rounded; marginal calluses thickened, slightly flanged; aperture proportionally narrow, straight, widening slightly at anterior end; outer lip edged with 26–30 long, thin teeth; columella with 28–30 long, thin teeth that extend into aperture and across columellar area; fossula present, proportionally small, deeply concave, lined with 12–13 large teeth; shell color pale tan-brown to orange-brown, overlaid with numerous small white spots; spots often blur together to form irregular long zebra stripes and amorphous patches; thickened anterior and posterior ends of shell white; margins and base of shell pale cream-tan or whitish-tan; apertural teeth reddish-brown; interior of aperture white or pale lavender-white.

Holotype: Length 53 mm, width 31 mm, LACM 3222; **Paratype:** length 50 mm, same locality as holotype, in the research collection of the author.

Type Locality: Under coral slabs in 2 m depth off Isla Levisa, Golfo de Guacayanabo, west of Manzanillo, Cuba.

Etymology: Named for Lindsey Groves, collections manager for the Department of Malacology at the Los Angeles County Museum of Natural History, in recognition of his many contributions to living and fossil cypraeids.

Discussion: This southern Cuban–northern Caribbean Sea subspecies has, for several years now, been referred to the taxon *Macrocypraea cervus peilei* (Schilder, 1932) by several cowrie workers such as Felix Lorenz and Marcus Coltro or was considered to be simply a dwarf form of the Carolinian province *M. cervus cervus* (Linnaeus, 1771). The true *M. cervus peilei* is an early Pleistocene fossil found in a quarry at Fort Catherina on Bermuda. The Bermudan fossil is smaller than typical living mainland American *M. cervus* and differs from the Cuban *M. cervus lindseyi* in being a more globose and less cylindrical shell with less projecting extremities. The new southern Cuban subspecies differs from the living Carolinian province *M. cervus* and its fossil forms (Petuch, 1994) in having a much smaller and heavier shell; in having much thicker and better developed lateral calluses; in having much thicker and more projecting anterior and posterior extremities; in having a lighter-colored shell, often with a white or pale whitish-violet aperture instead of dark purple as seen in *M. cervus*; in having fewer teeth on the outer lip (26–30 on *cervus lindseyi*; 40–50 on *cervus cervus*); and in having a proportionally much narrower and straighter aperture. In the latter character, *M. cervus lindseyi* shows a closer relationship to the widespread western Atlantic *M. zebra* (Linnaeus, 1758) than it does to the inflated, wide-apertured *L. cervus cervus* from the Gulf of Mexico, Florida, and the Carolinas. Unlike the other four *Macrocypraea* species (*cervinetta*, *zebra*, *zebra dissimilis*, and *cervus*), *M. cervus lindseyi* has prominent, distinct, bright white anterior and posterior extremities, a characteristic that is unique to the new subspecies.

Velutinoidea

Family Triviidae
Subfamily Triviinae

Genus *Pusula* Jousseaume, 1884

Pusula bessei new species
(Figure 7.11B,C)

Description: Shell large for genus, inflated, with high domed dorsum; margins thickened and flaring, base rounded; dorsum ornamented with 17 large, prominent, heavily beaded ribs, separated by wide, deep dorsal furrow; aperture thin and narrow; inner edge of lip with 12 large teeth, which correspond to large basal ribs; columella with 13 large teeth and corresponding ribs; basal ribs and apertural teeth sometimes bifurcating; fossula large and well developed, concave and depressed, with 6 large teeth; dorsum colored bright pinkish-tan or reddish-pink, with 3 pairs of large dark brown spots, with one pair at anterior end, one pair at midbody, and one pair at posterior end; dorsal spots and reddish-pink dorsal color separated by white dorsal furrow; beads on dorsal ribs lighter pink in color, speckled with pale brownish-pink; base and margins deep magenta-red, with lighter pinkish-red ribs; interior of aperture and fossula white.

Holotype: Length 18 mm, width 13.8 mm, FMNH 328408.

Type Locality: Found under coral slab in 3 m depth, off Coxon's Hole, Roatan Island, Bay Islands, Honduras.

Etymology: Named for Bruno Besse, formerly of Roatan Island, now of France, who collected the holotype.

Discussion: At 18 mm, *Pusula bessei* is one of the largest triviids found in the tropical western Atlantic. The new Roatan species is closest to the widespread western Atlantic *Pusula pediculus* (Linnaeus, 1758) but differs in being a more rounded and inflated shell with a proportionally higher, more domed dorsum; in having coarser and more heavily beaded dorsal ribs; in having proportionally smaller brown dorsal spots; and in being a far more colorful shell, with a bright reddish or pinkish-tan dorsum and white dorsal furrow and bright magenta base and margins. In this last character, *P. bessei* is also similar to the red-based *P. pacei* (Petuch, 1987) from the Bahamas but differs in being a much larger shell; in having a lighter-colored dorsum; and in having fewer and more strongly beaded dorsal ribs (Petuch, 1987). At present, the new triviid has only ever been collected on Roatan Island, to where it may be endemic.

Muricoidea

Family Muricidae
Subfamily Muricinae
Genus *Dermomurex* Monterosato, 1890

Dermomurex coonsorum new species
(Figure 6.13C,D)

Description: Shell small for genus, elongated, fusiform, with high, elevated spire; suture indented and spire whorls distinctly scalariform; shoulder rounded; body whorl and spire whorl with 3 large, rounded, winglike varices; elongated, rounded, bladelike longitudinal rib present between each pair of varices on body whorl; body whorl with 4 large, rounded primary spiral cords and single thinner secondary cord anterior of four primaries; spire whorls with 2 large spiral cords; siphonal canal narrow, well developed, ornamented with large, rounded spiral cord; intricalx chalky, microscopically pitted; aperture proportionally large, flaring, oval in shape; inner lip smooth; parietal shield proportionally large, adherent; shell uniformly cream-white.

Holotype: Length 10 mm, width 5 mm, FMNH 328409; **Paratype:** length 10 mm, same locality as holotype, in the research collection of the author.

Type Locality: In tide pool at low tide (with hermit crab), outer barrier reef of Southeast Cay, Glover's Atoll, Belize.

Etymology: Named for Charles ("Chuck") and Linda Coons, of Camano Island, Washington, who assisted the author in collecting and who found the holotype on Glover's Atoll.

Discussion: The new Belizean muricid, with its three wing-like varices and large blade-like intravarical ribs, is unlike any other known living western Atlantic *Dermomurex* species. Of the Atlantic members of its genus, *Dermomurex coonsorum* is most similar to *D. sarkini* E. Vokes, 1992, an early Pleistocene fossil species from the Moin Formation of Costa Rica. Although similar in size and general shape, *D. coonsorum* differs from the fossil *D. sarkini* in being a more elongated, narrower, and less inflated shell; in having a more rounded shoulder and varices; and in having only four primary spiral cords instead of five. At present, the new muricid is known only from Glover's Atoll, Belize, to where it may be endemic.

Subfamily Muricopsinae
Genus *Murexiella* Clench and Farfante, 1945

Murexiella deynzerorum new species
(Figure 6.7C,D)

Description: Shell of average size for genus, with globose, inflated body and high, stepped spire; shoulder sharply angled, with flattened subsutural area; whorls ornamented with 6–7 large, sharp varices per whorl and 5 large spiral cords; intersection of spiral cord and varix producing short recurved spine, giving shell rough appearance; spire whorls with 2 large spiral cords; siphonal canal elongated, ornamented with 3 large spiral cords; aperture proportionally large, round; parietal shield well developed, erect, not adherent, forming distinct peristome; shell color white with 2 wide red bands, one around edge of suture along subsutural area and one around midbody; some specimens (like paratype) deep rose-red color with 2 brown bands; interior of aperture white with bright red band.

Holotype: Length 21 mm, width 15 mm, FMNH 328410; **Paratype:** length 22 mm, same locality as holotype, in the research collection of the author.

Type Locality: In coral rubble near turtle grass beds, 3 m depth off Green Turtle Cay, Abaco Islands, Little Bahama Bank, Bahamas.

Etymology: Named for Albert ("Al"), Beverly ("Bev"), and Neal Deynzer, owners of Showcase Shells, Sanibel, Florida, and renowned divers and naturalists.

Discussion: This brightly colored, red-banded *Murexiella* most closely resembles a still unnamed species from Biscayne Bay, Florida, and the Florida Keys, which has been referred to by most authors as *Murexiella macgintyi* (M. Smith, 1938). The real *Murexiella macgintyi*, however, is a late Pliocene and early Pleistocene fossil from the Caloosahatchee formation of southern Florida and only superficially resembles the living Florida species (see Petuch, 1994). The Bahamian *M. deynzerorum* differs from the unnamed Florida species in being a much narrower, more slender shell with a more sharply angled shoulder and much longer siphonal canal; in having a higher and more protracted spire; in having much more stepped and scalariform spire whorls; in having proportionally smaller and less developed spines; and in having a distinctive color pattern composed of red bands.

Genus *Muricopsis* Bucquoy, Dautzenberg, and Dollfuss, 1882

Muricopsis honkeri new species
(Figure 6.7E,F)

Description: Shell of average size for genus, narrow, elongated and fusiform, with high, elevated spire; body whorl and spire with 6 narrow varices per whorl; each varix with 4 large, rounded spines, with posterior 3 being largest and anteriormost spine being smallest; 3–4 small scaly cords present between each pair of spines on body whorl; varices of spire whorls with one large spine per whorl; spire varices staggered, twisting in spiral pattern; siphonal canal short, narrow, with single large spine on each varix; aperture proportionally small, oval, with 5 large teeth along edge of lip; parietal shield adherent, ornamented with 2 small teeth at anterior end; shell color reddish-brown; varices of body whorl and spire whorls and spines white, producing overall zebra-like color pattern; interior of aperture and parietal shield pinkish-white.

Holotype: Length 21 mm, width 7 mm, FMNH 328411; **Paratype:** length 20 mm, in the research collection of the author.

Type Locality: Found in coral rubble in 3 m depth off Green Turtle Cay, Abaco Islands, Little Bahama Bank, Bahamas.

Etymology: Named for Thomas Honker of Delray Beach, Florida, noted diver and shell dealer, who collected the type material on Green Turtle Cay.

Discussion: This beautiful new Bahamian muricid is most similar to the widespread Caribbean *Muricopsis oxytatus* (M. Smith, 1938) but differs in being a much smaller, stockier shell with a proportionally lower spire; in having shorter, broader, and flatter varical spines; in having only one spine on the siphonal canal; and in having the distinctive zebra-like color pattern of a reddish-brown body whorl and white varices. In this last feature, *Muricopsis honkeri* also resembles *Muricopsis chesleri* Houart, 2006 from Roatan Island, Honduras, but differs in having spirally twisting spire varices instead of straight ones and in having a reddish-brown base color instead of black (as in *chesleri*). The new species appears to be endemic to the Abaco Islands.

Subfamily Ergalitaxinae

Roquesia new genus

Diagnosis: Shells very small for family and subfamily, tiny, elongated and fusiform; body whorl inflated, more rounded than spire whorls; shoulder rounded, blending into body whorl; body whorl ornamented with 26 thin, evenly spaced longitudinal cordlike riblets which, in turn, are crossed by 20 thin spiral cords, producing strong cancellate sculpture; body whorl and spire whorls with 2 large, fin-like varices per whorl; varices staggered, not aligned, squared off along posterior edges; siphonal canal well developed, open, ornamented with 6 large spiral cords; protoconch proportionally very large, rounded, dome-like, projecting, composed of 2 whorls; aperture large, oval, bordered by large, thickened varix along lip; columellar area smooth, with small, nearly obsolete parietal shield; shells colored pale cream-white.

Type Species: *Roquesia lindae* new species, by monotypy. From 2 m depth under coral slab, off the southern shore of Crasqui Island, Los Roques Atoll, Venezuela.

Etymology: Named for the Los Roques Atoll, the type locality and the only known area that the genus has been collected.

Discussion: This new genus of tiny ergalitaxine muricids is most similar to the genus *Minibraria* Sarasua, 1984 from the northern Caribbean and Gulf of Mexico (as typified by *Minibraria monroei*; McGinty, 1962) but differs in having a more inflated body whorl; in having proportionally larger and more pronounced varices; and in having stronger and more pronounced cancellate sculpture. In this last shell character, *Roquesia* is most similar to members of the Eastern Pacific (Panamic Province) and South Pacific genus *Phyllocoma* Tapparone-Canefri, 1881, and, indeed, somewhat resembles a tiny specimen of the Panamic *Phyllocoma scalariformis* (Broderip, 1833). Besides its tiny size, *Roquesia* differs from *Phyllocoma* in lacking prominent opposing varices on the body whorl and in lacking the large, flaring parietal shield. At present, *Roquesia* is only known from the Los Roques Atoll, to where it may prove to be endemic.

Roquesia lindae new species
(Figure 9.6A,B)

Description: General shell morphology as for genus; unique holotype with single large brown patch on lateral area of dorsum; interior of aperture cream-white.

Holotype: Length 9 mm, width 4 mm, FMNH 328412.

Type Locality: In 2 m depth, under coral slab, off southern end of Crasqui, Los Roques Atoll, Venezuela.

Etymology: Named for Linda J. Petuch, wife of the author.

Discussion: The phylogenetic relationships of this new ergalitaxine muricid were discussed in the previous section. The unique holotype was collected dead, occupied by a hermit crab, in an area of strong currents and wave surge.

Subfamily Rapaninae
Genus *Stramonita* Schumacher, 1817

Stramonita buchecki new species
(Figure 3.6A,B)

Description: Shell of average size for subgenus, inflated, with elevated, scalariform spire; shoulder sharply angled, with broadly sloping subsutural area; spire whorls acuminate, with slightly concave profiles; body whorl ornamented with 3 large, nodulose spiral cords, with the largest 2 cords being around the shoulder and midbody; smaller nodulose spiral cord around junction of siphonal canal and body whorl; shoulder and spire whorls ornamented with 10 large, slightly flattened knobs per whorl; smaller spiral cords present between primary sculpture of strong cords and knobs, with 8 small cords present on the sloping subsutural area of the shoulder and sets of 4 small cords between the 3 large knobbed cords on the body whorl; shell color pale orange-white or whitish-tan, overlaid with variable amounts of pale brown and brownish-gray flammules and patches; knobbed cords around midbody marked with large, evenly spaced whitish-tan patches; rows of small brown dots and dashes present on anterior half of shell; siphonal canal short, open, colored a light orange; aperture proportionally large, wide, flaring, colored bright orange; edge of lip highly crenulate, with inner edge of lip baring 6 large, elongated rib-like teeth; parietal area and inside of siphonal canal bright orange; anal ridge present along posterior of aperture, colored pale orange-white.

Holotype: Length 28 mm, width 16 mm, FMNH 328413; **Paratype:** length 28 mm, same locality as the holotype, in the research collection of the author.

Type Locality: Under an oyster-covered rock at low tide, Pine Point, Singer Island, Lake Worth Lagoon, Riviera Beach, Palm Beach County, Florida.

Etymology: Named for Joseph Bucheck II, of Pine Point, Singer Island, who assisted the author in his research in the Everglades fossil beds, the Florida Keys, and Lake Worth.

Discussion: This new Palm Beach rapanine muricid closely resembles the related *Stramonita rustica* (Lamarck, 1822) (Figure 3.6C) but differs in having a much more inflated and proportionally wider body whorl; in being much broader across the shoulder; in having proportionally larger and flatter shoulder knobs; in having more acute, narrower, and distinctly concave spire whorls; in having a pale orange base shell color; and, most importantly, in having a bright orange aperture, outer lip, and parietal area (as opposed to white in *S. rustica*). Both species have the same type and number of large rib-like teeth on the outer lip and have similar aperture shapes. At Pine Point, Singer Island, both *Stramonita buchecki* and *Stramonita rustica* occur together on the oyster-covered rocks at low tide but can always be differentiated by the white aperture color of *S. rustica* and the bright orange color of the new species. Although having the same type of sculpture and being the same size as *S. rustica*, *S. buchecki* is consistently a more inflated shell with a more acute spire. In having an orange aperture, the new species also somewhat resembles the common *Stramonita floridana* (Conrad, 1837) but differs in being a smaller, more heavily knobbed shell with a narrower, more scalariform spire; in having the 6 large teeth inside the outer lip (instead of the smooth inner lip of *S. floridana*); and in having a more intense orange apertural color. All three *Stramonita* species occur together at Pine Point, with *S. floridana* being the most common species and *S. buchecki* being the rarest.

The *Stramonita* cf. *rustica* shown in the *Encyclopedia of Texas Sea Shells* (Tunnel et al., 2010: 212) is similar to *S. buchecki* and may represent the same species or, most likely, another new species endemic to the Texas coast of the northwestern Gulf of Mexico. The unnamed Texas species has the same inflated body whorl and large knobs but lacks the intense bright orange apertural color and also the 6 large teeth on the inside of the outer lip. The discovery of a new *Stramonita* species in Lake Worth underscores the still unexplored nature of the Palm Beach Provinciatone.

Buccinoidea

Family Buccinidae
Subfamily Pisaniinae
Genus *Engina* Gray, 1839

Engina dicksoni new species
(Figure 5.6E)

Description: Shell of average size for genus, very elongated, fusiform, with high, protracted spire; shoulder angled, thickened; subsutural areas flattened, slightly sloping; shoulder ornamented with 8 large, rib-like knobs per whorl; knobs extend onto body whorl as low, undulating longitudinal ribs; body whorl sculptured with 12 large spiral cords, which become stronger and coarser on anterior end; spire whorls ornamented with 4 large spiral ribs; shell color pale orange-cream or salmon-orange, overlaid with 2 darker orange bands, one around shoulder and one around mid-body; coarse spiral ribs around anterior end marked with evenly spaced pale orange spots; aperture wide and flaring, with interior of aperture and columella being white; structure of apertural dentition unknown (since the type specimens are all subadults); protoconch pale orange in color, proportionally large and mamillate, composed of 2 whorls; siphonal canal short, open.

Holotype: Length 18 mm, width 9 mm, FMNH 328420; **Paratype:** length 19 mm, same locality as the holotype, in the research collection of the author.

Type Locality: Trawled by commercial shrimpers from 35 m depth on Campeche Bank, off Progresso, Yucatan State, Mexico.

Etymology: Named for Andrew Dickson of Miami, Florida, who collected the type material and generously donated it for research.

Discussion: This new Yucatanean species is most similar to *Engina corinnae* Crovo, 1971 from the southeastern Florida (Floridian Subprovince) but differs in being a larger, much more elongated shell with a proportionally much higher spire and in having a color pattern of pale orange-cream with orange bands instead of white with large brown patches (as in *corinnae*). Since both type specimens (the only known specimens) are subadults with thin outer lips, the structure of the adult apertural dentition is still unknown at this time.

Family Fasciolariidae
Subfamily Fasciolariinae

Genus *Cinctura* Hollister, 1957

Cinctura hunteria keatonorum new subspecies
(Figure 3.6D)

Description: Shell of average size for species, elongated, with rounded whorls and sloping shoulder; spire elevated, protracted; siphonal canal short, open; shell color pale salmon-white to yellowish-cream with numerous irregular, narrow, longitudinal salmon-orange or orange-tan flammules; flammules best developed along subsutural area and around midbody; base color pattern overlaid with 6 thin, regularly spaced dark brown lines; thinner and less developed secondary lines frequently present between primary lines; siphonal canal darker orange-tan or salmon-orange; aperture oval, with white interior; parietal area with pale salmon-orange glaze.

Holotype: Length 92 mm (incomplete, protoconch missing), width 43 mm, FMNH 328415; **Paratypes:** length 91 mm, same locality as the holotype, in the research collection of the author; length 79 mm, Keaton collection.

Type Locality: Dredged by a commercial scallop boat from 35 m depth off Cape Canaveral, Brevard County, Florida.

Etymology: Named for Dr. Kenneth and Mrs. Wendy Keaton of Lauderhill, Florida, who collected the type material at the Canaveral scallop cleaning station (now closed).

Discussion: This new offshore, deep-water subspecies of the shallow water, intertidal *Cinctura hunteria* (Perry, 1811) differs from the nominate subspecies in having consistently more slender and elongated shell with proportionally higher spires; in being colored pale salmon-orange with darker salmon-orange and orange-tan longitudinal flammules; and in having salmon-orange parietal areas and siphonal canals. The widespread shallow water nominate subspecies, *Cinctura hunteria*, is always a more inflated, stockier shell and is most frequently colored a dark khaki-green mixed with brown and dark orange-tan. The new subspecies, *Cinctura hunteria keatonorum*, ranges from Cape Hatteras, North Carolina, south to the Cape Canaveral area of Florida, where it lives in deeper water areas (30–50 m) in offshore scallop beds. This new subspecies may eventually prove to be a full species, an ecological sibling of its shallow water relative.

The discovery of a new deep-water subspecies of *Cinctura hunteria* shows that the *Cinctura* complex of the Carolinian province was larger than originally thought, comprising five distinct species and subspecies: (1) *Cinctura hunteria* (Perry, 1811), the widest-ranging species, which is found in shallow intertidal areas from Cape Hatteras, North Carolina, south to Florida, throughout the Florida Keys, and westward to Texas; (2) *Cinctura hunteria keatonorum* Petuch and Sargent, 2012, which lives in offshore scallop beds from Cape Hatteras to Cape Canaveral; (3) *Cinctura tortugana* Hoolister, 1957, which occurs in deeper water areas (25–100 m depths) off the Florida Keys and Dry Tortugas and offshore of southwestern Florida; (4) *Cinctura lilium* Fischer, 1807, which lives in offshore areas (20–100 m depths) from Louisiana to Veracruz, Mexico; and (5) *Cinctura branhamae* Rehder and Abbott, 1951, which occurs on the Campeche Banks and offshore of the Yucatan Peninsula of Mexico. Although placed in *Cinctura* by Abbott (1974), the large, deep-water *Fasciolaria bullisi* Lyons, 1972 (from off western Florida and the Yucatan Peninsula) is now known to be a member of *Fasciolaria* (s.s.) and does not belong to the *Cinctura* species complex (William Lyons, personal communication).

Subfamily Peristerniinae
Subgenus *Polygona* Schumacher, 1817

Polygona bessei new species
(Figure 7.11G)

Description: Shell small for genus, thin, delicate, very elongated, with high protracted spire and narrow, extended siphonal canal; spire of adult with 7 scalariform whorls; spire whorls and body whorl rounded, with 7 large, narrow, varix-like knobs per whorl; body whorl sculptured with 14–15 low, thick primary cords; thin, threadlike smaller secondary cords sometimes present between larger primary cords, producing overall rough texture; shell color uniform deep orange-brown; aperture proportionally small, oval in shape, pinkish-white on interior, with 10 large cords on inside of lip; large beadlike knobs present at anal notch at posterior end and on columella at anterior of body whorl; columella with 3 low plicae; parietal shield thin, adherent; protoconch proportionally very large, rounded, domelike, composed of 2 whorls.

Holotype: Length 33 mm, width 12 mm, FMNH 328416; **Paratype:** length 32 mm, same locality as holotype, in the research collection of the author.

Type Locality: Brought up in commercial lobster traps, from 250 m depth off Coxon's Hole, Roatan Island, Bay Islands, Honduras.

Etymology: Named for Bruno Besse, formerly of Coxon's Hole, Roatan Island, now living in France, who collected the type material.

Discussion: This new deep-water Roatan species most closely resembles *Polygona ceramidus* (Dall, 1889), another deep-water species that is endemic to Barbados. Although similar in overall shape and size, *Polygona bessei* differs from its Barbados congener in being a much more slender and elongated species with a proportionally higher spire. The shoulders of the body and spire whorls of *P. ceramidus* are proportionally broader than those of the new Roatan species and are sharply angled instead of being rounded in outline. At present, *Polygona bessei* is known only from the deep-water areas around the Bay Islands.

Polygona paulae new species
(Figure 6.7G)

Description: Shell of average size for genus, with rounded, inflated whorls; spire high, protracted, distinctly stepped, of equal size to body whorl; shoulder rounded; body and spire with 8 large, highly raised, thickened varices per whorl; body whorl ornamented with 14 large, prominent, raised spiral cords; entire shell ornamented with numerous thin, raised, longitudinal squamous growth lines, giving shell scaly appearance; suture bordered with 2 very thin spiral cords; siphonal canal proportionally long, narrow, ornamented with 3 large spiral cords and several smaller secondary spiral cords; aperture proportionally large, flaring, oval; edge of outer lip with 13–14 large beaded riblike teeth; columella with small, thin parietal shield; parietal shield adherent, ornamented with

4 large teeth on anterior half; narrow and deep pseudoumbilicus present at anterior end of columellar area; shell color rich orange-brown, with spiral cords being darker brown; anterior tip and sutural area dark brown; interior of aperture light orange.

Holotype: Length 48 mm, width 25 mm, FMNH 328417; **Paratype:** length 35 mm, same locality as the holotype, in the research collection of the author.

Type Locality: Under coral rubble, 3 m depth on reef off Green Turtle Cay, Abaco Islands, Little Bahama Bank, Bahamas.

Etymology: Named for Paula Honker of Delray Beach, Florida, wife of Thomas Honker and inspired amateur naturalist and diver, who collected the type material in the Abacos.

Discussion: This new species of Bahamian *Polygona* is most similar to the widespread Caribbean *Polygona angulatus* (Roeding, 1798) but differs in having much broader, more inflated whorls; in having more indented sutures on the spire whorls; in having a much more stepped, scalariform spire; in having fewer, much coarser, and more prominent spiral cords; and in having a proportionally much larger and more flaring aperture. By having widely spaced prominent brown spiral cords, *Polygona paulae* also resembles the widespread Caribbean *Polygona infundibulum* (Gmelin, 1791) but differs in being a less elongated, much stockier, and more inflated shell with a lower spire. To date, the species has been collected only in the Abacos.

Family Busyconidae
Subfamily Busycotypinae
Genus *Fulguropsis* Marks, 1950

Fulguropsis spiratum keysensis new subspecies
(Figure 4.4E)

Description: Shell small for species, heavy and thickened, distinctly pyriform; shoulder sharply angled, bordered by strong raised carina; subsutural area sloping on body whorl, flatter and slightly depressed or canaliculate on spire whorls; spire high, protracted, elevated, distinctly stepped and scalariform; suture bordered by narrow, deep channel; shell heavily ornamented with numerous strong spiral cords, which are strongest on shoulder and around siphonal canal; shell colored pale bluish-gray and yellow-cream overlaid with regularly spaced dark brown amorphous longitudinal flammules, which extend onto spire whorls; interior of aperture heavily ornamented with numerous strong lirae, colored deep brown within; inner edge of lip and siphonal canal white with few small brown patches along edge; protoconch pale tan, proportionally large and bulbous, composed of 1.5 whorls.

Holotype: Length 57.8 mm, width 34 mm, FMNH 328418.

Type Locality: Found in a sand pocket on a sponge bioherm, 1.5 m depth off the northeastern end of Little Torch Key, Lower Florida Keys, Monroe County, Florida.

Etymology: Named for the Florida Keys, the area to where the new subspecies is restricted.

Discussion: The common pear whelk, *Fulguropsis spiratum pyruloides* (Say, 1822), which is found in shallow coastal waters in the Carolinas, Georgia, and both sides of the Floridian Peninsula, has never been reported from the coral reef areas of the Florida Keys reef tract or the shallow carbonate sand areas of the lower Keys. Recent investigations of the sponge bioherms and limestone sea floor areas off Little Torch Key (Petuch and Sargent, 2012), however, have uncovered the first *Fulguropsis* known from that area. Closer examination showed that this dwarf busyconid constituted a new subspecies, in many ways more closely resembling Pleistocene fossil species than it does the living *F. spiratum pyruloides*. The new endemic lower Keys subspecies, *Fuluropsis spiratum keysensis*, differs from the more northerly *pyruloides* by being a much smaller, narrower, more cylindrical, and far less inflated shell; in having an elevated, protracted spire, with distinctly stepped and scalariform spire whorls; and in having prominent, very strong lirae within the aperture (*pyruloides* is always smooth within the aperture and lacks any lirae). In this last character, *F. spiratum keysensis* is similar to *F. spiratum spiratum* (Lamarck, 1816) from the Louisiana, Texas, and Mexico coasts. That nominate subspecies, although having strong apertural lirae, has a rounded, inflated body whorl with a flattened spire and more closely resembles the eastern Carolinian subspecies *pyruloides*. The new Keys subspecies also differs from both *spiratum* and *pyruloides* in having a greatly reduced, much shallower, and narrower sutural channel on the spire whorls, a shell character very much in common with the Pleistocene fossil *F. spiratum pahayokee* Petuch, 1994 from the Everglades area.

Volutoidea

Family Turbinellidae
Subfamily Vasinae
Genus *Globivasum* Abbott, 1950

Globivasum globulus whicheri new subspecies
(Figure 9.6C,D)

Description: Shell of average size for genus and species, biconic, with high, elevated, stepped spire; body whorl inflated, rotund, with 8 large, rounded varices per whorl; shoulder sharply angled, edged with large spiral cord and 8 large spines that correspond to varices; subsutural area distinctly sloping, slightly concave; body whorl ornamented with 3 large spiral cords; 3 or 4 small scaly secondary cords present between each pair of large primary cords; spire whorls ornamented with only 6 fine, scaly secondary cords; siphonal canal short, contracted, distinctly separated from rest of body whorl; siphonal canal ornamented with 6–8 fine scaly secondary cords and single cord with large rounded spines; aperture elongated, oval, with 12 very low, rounded, elongated teeth on inner edge of lip; columella with 3 large plications; small open pseudoumbilicus present; shell color pale tan-brown with white primary ribs and shoulder spines; aperture yellow, becoming deeper golden-orange farther in; columella and columellar plications golden-yellow.

Holotype: Length 29 mm, width 24 mm, FMNH 328423; **Paratype:** length 28 mm, same locality as the holotype, in the research collection of the author.

Type Locality: Found on open limestone rock bottom in strong currents, 3 m depth off the southeastern side of Prickly Pear Cays, Anguilla Island, British West Indies, Lesser Antilles.

Etymology: Named for Dr. John Whicher of Somerset, England, who collected the type specimens off Anguilla.

Discussion: As presently understood, the unique Caribbean province (Grenadian Subprovince) genus *Globivasum* is represented by two geographically isolated populations; one in the north on Anguilla Island and the neighboring Prickly Pear Cays (*Globivasum globulus whicheri* new subspecies) and the other in the south on Barbuda and Antigua Islands (*Globivasum globulus*; Lamarck, 1816). The two subspecies are separated by over 150 km distance and occur on two different submarine platforms that are separated by deep-water channels. Being direct-development animals with nonplanktonic larvae, these two small vase shells have become bathymetrically and geographically isolated from each other and are undergoing genetic drift, leading to the evolution of different shell morphologies. The Anguillan *Globivasum globulus whicheri* differs from the nominate subspecies from Antigua and Barbuda in having a more elongated shell with a much higher, more protracted spire; in having a sharply angled shoulder bordered with large spines; and in having a better developed and narrower siphonal canal. The nominate subspecies, *G. globulus globulus*, on the other hand, has a very rounded shape with no distinct shoulder and has a very low, only slightly raised spire (Figures 9.6E,F). Because of its high spire, large shoulder spines, and longer siphonal canal, *G. globulus whicheri* is sometimes confused with the larger, more elongated *Vasum capitellum* (Linnaeus, 1758). Future research may show that the two subspecies of *G. globulus* are actually distinct, but closely related, species.

Family Volutidae
Subfamily Lyriinae
Genus *Enaeta* H. and A. Adams, 1853

Enaeta bessei new species
(Figure 7.11D,E)

Description: Shell small for genus, stocky and rotund, broad across shoulder, smooth and silky; spire subpyramidal, truncated; shoulder rounded, only slightly angled, with sloping, slightly indented subsutural areas; body whorl and spire whorls ornamented with 14–15 large, evenly spaced, thin longitudinal ribs; low, elongated rounded knobs present on shoulder, corresponding to longitudinal ribs; 2 very thin spiral threads present around subsutural area; shell color pale cream-white, with thin discontinuous band of reddish-brown flammules around midbody; outer lip expanded into large rounded terminal varix; aperture proportionally elongated and narrow, white on interior; inner edge of lip with 5 very small, low, almost obsolete teeth; columellar area with 5 large, thin teeth that extend onto parietal shield; protoconch proportionally large, dome-like; early whorls and protoconch colored reddish-brown.

Holotype: Length 10 mm, width 5.5 mm, FMNH 328424.

Type Locality: Collected by commercial lobster divers, from 10 m depth in coral rubble, on Rosalind Bank, Honduras, Central Caribbean Banks.

Etymology: Named for Bruno Besse, formerly of Roatan Island, Honduras, now of France, who donated the holotype to the author.

Discussion: At 10 mm in length, *Enaeta bessei* is the smallest-known volute found in the tropical western Atlantic. With its stocky, rotund shape, this new Honduran species is most similar to the Bahamian *E. cylleniformis* (Sowerby, 1844) but differs in being a much smoother and less sculptured shell that lacks spiral threads and cords; in having a much lower and distinctly truncated spire; in having a more rounded and less angled shoulder; in having a brown protoconch and early whorls; and in lacking the single large labial tooth in the aperture. The new species appears to be endemic to the isolated Rosalind Bank.

Enaeta lindae new species
(Figure 6.8A,B)

Description: Shell small for genus, with dull, silky texture, very elongated, with high, elevated spire; shoulder angled, with angle being more distinct on spire whorls; body whorl ornamented with 14–15 thin, prominent costae; costae of first half of body whorl and of spire whorls ornamented with small, smooth, elongated bead-like knob at shoulder angle; anterior end ornamented with numerous very fine spiral threads; shell color pale cream-white with scattered large dark brown amorphous flammules; outer lip thickened and enlarged, producing prominent rounded terminal varix; aperture elongately oval, with 7–8 very faint, tiny teeth along inner edge of lip; lip with single large, prominent knob-like tooth just posterior to shell midline; columellar area with 4 large, elongated teeth that extend onto parietal shield; interior of aperture white; protoconch proportionally very large, rounded, dome-like, composed of 1.5 whorls.

Holotype: Length 13 mm, width 5 mm, FMNH 328425; **Paratype:** length 13 mm, same locality as the holotype, in the research collection of the author.

Type Locality: Under coral slab in 5 m depth, on reef off "Tuna Alley," western side of Cat Cay, southern Bimini Chain, Great Bahama Bank, Bahamas.

Etymology: Named for my wife, Linda Joyce Petuch.

Discussion: This small lyriine volute occurs in the same general locality as the classic Bahamian volute *Enaeta cylleniformis* (Sowerby, 1844) but differs in being a much smaller, being a much more elongated and cylindrical shell with a far less inflated body whorl and in having a more colorful shell, with large dark brown amorphous flammules and patches. In general shell shape and size, *Enaeta lindae* strongly resembles the Honduran and Central Caribbean *E. reevei* (Dall, 1907) but differs in being a more slender and cylindrical shell; in having less indented areas around the suture; in having fewer and thicker ribs on the body whorl and spire (14–15 in *lindae* and 18–19 on *reevei*); and in being a more colorful shell, with large dark brown patches. The new Bahamian species is the most recently discovered member of a large species complex that comprises at least eight species:

> *Enaeta bessei* new species (described previously; Rosalind Bank, Honduras)
> *Enaeta cylleniformis* (Sowerby, 1844) (Bahamas and Biscayne Bay, Florida)
> *Enaeta guildingi* (Sowerby, 1844) (Lesser Antilles and the Grenadines)
> *Enaeta leonardhilli* Petuch, 1988 (Fernando de Noronha Island, Brazil)
> *Enaeta lindae* new species (Bimini Chain, Bahamas)
> *Enaeta reevei* (Dall, 1907) (southern Cuba, Honduras, and Central Caribbean Banks)

Enaeta new species a. (coastal Brazil)
Enaeta new species b. (Biscayne Bay and upper Florida Keys, Florida)

Another small *Enaeta*, resembling *E. reevei*, has also been collected in the San Blas Islands of Panama and may represent another new species.

Family Olividae
Subfamily Olivinae
Genus *Oliva* Bruguiere, 1789

Americoliva new subgenus

Diagnosis: Shells large for genus *Oliva*, elongated, generally fusiform, with straight or slightly convex sides; shells highly polished, glossy; shoulders nonexistent, with body whorls blending directly into spires; spires generally elevated, subpyramidal, with some species having somewhat scalariform spires and others having low, pyramidal spires; edge of suture deeply canaliculate, with thin spiral groove; spire whorls enameled; apertures narrow, becoming wider toward anterior end; columellas well developed, lined with numerous large, tooth-like plications; plicae become larger toward anterior end; shell colors generally white, pale pink, pale blue, or cream-tan, overlaid with variable amounts of large, tan or brown triangular markings arranged in a net-like pattern; two bands of darker triangular markings or amorphous flammules almost always present around midbody; edge of spire suture characteristically marked with large purple-brown or reddish-brown elongated longitudinal flammules, overlaid with thin brown longitudinal hairlines; dark suture flammules separated by evenly spaced white triangular markings along edge of suture channel; protoconchs proportionally large, rounded, bulbous, composed of 2 or 3 whorls.

Type Species: *Oliva sayana* Ravenel, 1834, Carolinian province, from North Carolina to Florida and the Gulf of Mexico.

Other Species in *Americoliva*: (Western Atlantic Species) *Oliva (Americoliva) antillensis* Petuch and Sargent, 1986, Caribbean Sea; *Oliva (Americoliva) bahamasensis* Petuch and Sargent, 1986, deep water off the Bahamas; *Oliva (Americoliva) barbadensis* Petuch and Sargent, 1986, deep water off Barbados; *Oliva (Americoliva) bayeri* Petuch, 2001, Columbia; *Oliva (Americoliva) bewleyi* Marrat, 1870, northern South America; *Oliva (Americoliva) bifasciata* Kuester, 1878, Caribbean Sea; *Oliva (Americoliva) bifasciata bollingi* Clench, 1937, Carolinian Province; *Oliva bifasciata sunderlundi* Petuch, 1987, deep water off western Florida; *Oliva (Americoliva) bifasciata jenseni* Petuch and Sargent, 1986, Bermuda; *Oliva (Americoliva) broderipi* Ducros de St. Germaine, 1857, Hispaniola; *Oliva (Americoliva) broderipi* unnamed subspecies, Puerto Rico; *Oliva (Americoliva) circinata* Marrat, 1871, central Brazil; *Oliva (Americoliva) circinata jorioi* n. subsp. (described next), northern Brazil; *Oliva circinata tostesi* Petuch, 1987, southern Brazil; *Oliva (Americoliva) contoyensis* Petuch, 1988, Yucatan, Mexico; *Oliva (Americoliva) drangai* Schwengel, 1951, deep water off Tobago; *Oliva (Americoliva) fulgurator* Roeding, 1798, Venezuela; *Oliva (Americoliva) fulgurator bullata* Marrat, 1871, Los Roques Atoll and Las Aves Islands; *Oliva (Americoliva) fulgurator fusiforme* Lamarck, 1811, Aruba; *Oliva (Americoliva) goajira* Petuch and Sargent, 1986, deep water off Colombia; *Oliva (Americoliva) jamaicensis* Marrat, 1870, Jamaica; *Oliva (Americoliva) maya* Petuch and Sargent, 1986, Yucatan Peninsula; *Oliva (Americoliva) mooreana* new species (described next), deep water off Belize; *Oliva (Americoliva) olivacea* Marrat, 1870, Lesser Antilles; *Oliva (Americoliva) porcea* Marrat, 1870, Venezuela and Colombia; *Oliva (Americoliva) reclusa* Marrat, 1871, Aruba; *Oliva (Americoliva) reticularis* Lamarck, 1811, Caribbean Sea; *Oliva (Americoliva reticularis ernesti* Petuch, 1991, Panama; *Oliva (Americoliva) sayana sarasotaensis* Petuch and Sargent, 1986, deep water off western Florida; *Oliva (Americoliva) sayana texana* Petuch and Sargent, 1986, Texas. (**Eastern Pacific Species**) *Oliva (Americoliva) corteziana* Petuch and Sargent, 1986, Gulf of California; *Oliva (Americoliva) cumingii* Reeve, 1830, Gulf of California; *Oliva (Americoliva) ionopsis* Berry, 1969, Gulf of California; *Oliva (Americoliva) olssoni* Petuch and Sargent, 1986, Costa Rica and Panama; *Oliva (Americoliva) pindarina* Duclos, 1835, Gulf of California; *Oliva (Americoliva) polpasta* Duclos, 1835, Mexico, western Central America, to Ecuador; *Oliva (Americoliva) polpasta davisae* Durham, 1950, deep water in the Gulf of California; *Oliva (Americoliva) polpasta radix* Petuch and Sargent, 1986, deep water off Panama, Colombia, and Ecuador; *Oliva (Americoliva) rejecta* Burch and Burch, 1962, Gulf of California; *Oliva (Americoliva) spicata* Röding, 1798, west Central America, Costa Rica, Panama; *Oliva (Americoliva) spicata deynzerae* Petuch and Sargent, 1986, Cocos Island; *Oliva (Americoliva) truncata* Marrat, 1867, Costa Rica and Panama; *Oliva (Americoliva) venulata* Lamarck, 1811, Gulf of California; *Oliva (Americoliva) violacea* Marrat, 1867, Gulf of California.

Etymology: Named as a combination of America and *Oliva*, referring to the new genus being endemic to the tropical Americas.

Discussion: Previously (Petuch and Sargent, 1986: 119), this large American species complex had been placed in the subgenus *Strephona* Mörch, 1852 (Type: *Oliva flammulata* Lamarck, 1811), primarily because of the similarity in shell shapes and the presence of triangular "tent markings" on both groups. The American complex, however, differs from the West African *Strephona* species (*O. flammulata* from Senegal to Angola, and *O. verdensis* Petuch and Sargent, 1986, from the Cape Verde Islands) in having larger, more elongated shells and in having the band of large chevron-like flammules and hairlines along the edge of the filament channel. This characteristic band of flammules is found on all of the *Americoliva* species, both western Atlantic and eastern Pacific, but is absent on the West African *Strephona* species.

Oliva (Americoliva) circinata jorioi new subspecies
(Figure 10.3F)

Description: Shell small for subgenus, bluntly cylindrical, with straight sides; shoulder rounded; spire low and subpyramidal; shell color cream-white overlaid with network of proportionally very large, light brown and tan triangles; 2 bands of smaller triangles and scattered large brown dots present around midbody; filament channel edged with large purple chevron flammules and dark brown hairlines; spire whorls partially covered by pale tan-orange callus; aperture proportionally narrow, flaring slightly at anterior end, white within; columella with 16–18 large, thin plicae; protoconch proportionally large, cream-white in color, composed of 2 whorls.

Holotype: Length 32 mm, width 13 mm, FMNH 328428; **Paratype:** length 27 mm, same locality as holotype, in the collection of the author.

Type Locality: In sand, 10 m depth off Cabo de São Roque, Rio Grande do Norte State, Brazil.

Etymology: Named for Domingos A. ("Afonso") Jorio of Guarapari, Espiritu Santo State, Brazil, diver and naturalist, in recognition of his many contributions to Brazilian malacology.

Discussion: This new Cearaian Subprovince subspecies of the Bahian Subprovince *Oliva (Americoliva) circinata* Marrat, 1871 (shown in Figure 2.7 in this volume) differs from its southern nominate subspecies in being a much smaller, stockier, and less elongated shell; in having a proportionally lower spire; and in being much paler in color, with a white base color and pale tan triangle as opposed to the deep yellow base color and dark brown and blue-green triangles seen on *O. (Americoliva) circinata*. Characteristically, the triangle markings on *O. (Americoliva) circinata jorioi* are also proportionally much larger and more open, while those of *O. (Americoliva) circinata circinata* are proportionally much smaller and more closely packed.

Oliva (Americoliva) mooreana new species
(Figure 6.13E,F)

Description: Shell very small for subgenus, glossy, thickened and heavy, stocky and barrel-shaped, widest around midbody; spire subpyramidal, composed of only 3 whorls; aperture narrow; mantle filament channel proportionally deep and well developed; columella with 12 large, blunt, elongated teeth; fasciole proportionally small and confined to extreme anterior end; shell color pale salmon pink overlaid with 2 thin bands of tiny darker salmon-tan dots and flammules, one above midbody line and one below midbody line; edge of filament channel marked with 6 large, widely scattered darker orange-tan flammules per whorl; interior of aperture white; protoconch pale salmon-cream, proportionally extremely large, dome-shaped, composed of 1.5 large bulbous whorls.

Holotype: Length 20 mm, width 9 mm, FMNH 328429.

Type Locality: Dredged from 310 m depth off the Turneffe Islands, Belize.

Etymology: Named for Dennis and Penelope Ann ("Penny") Moore of Prospect, Oregon, and Loreto, Mexico, who assisted the author while collecting on Glover's Atoll, Belize.

Discussion: This small, stocky dwarf species is the smallest known member of the subgenus *Americoliva*. The thick, heavy shell of the holotype, along with its thickened, strong lip, demonstrates that the type specimen represents a fully adult shell. In having a salmon-pink shell color and in being short and fusiform in shape, the new species is similar to *Oliva (Americoliva) bifasciata sunderlandi* Petuch, 1987 from deep water (150 m depth) off western Florida (see Petuch, 1987). The new Belizean olive differs from the eastern Gulf of Mexico species in having larger, fewer, and more widely spaced flammules along the filament channel; in lacking rows of fine hairlines along the filament channel; in having a proportionally much larger and more bulbous protoconch; and in having a darker and brighter pink shell color. Having been collected in 310 m depth, *Oliva (Americoliva) mooreana* is the second deepest dwelling *Americoliva* in the western Atlantic and eastern Pacific, with only *Oliva (Americoliva) bahamasensis* Petuch and Sargent, 1986 coming from deeper water (350–400 m depth off the western Bahamas). The holotype of the new Belizean species was originally dredged along with the spectacular winged muricid *Pterynotus rad-wini* Harasewych and Jensen, 1979 and the turbinid *Bolma (Senobolma) sunderlandi* Petuch, 1987, showing that it is part of a unique deep-water ecosystem that occurs only along the Belizean coast (Petuch, 1987: 61,79).

Superfamily Conoidea
Family Conidae
Subfamily Puncticulinae

Arubaconus new genus

Diagnosis: Shell small for subfamily, obese and inflated, distinctly ovate in shape; spire low, turbinate, with indented sutures and rounded whorls; shoulder rounded, sloping, blending directly into body whorl; body whorl ornamented with 13–15 evenly spaced, large, beaded spiral cords; siphonal canal short, truncated, ornamented with 6–8 spiral cords; body whorl colored black or dark blackish-brown, with variable amounts of large, amorphous white flammules, often arranged in two main bands; anterior end purple; spire whorls white with evenly spaced thin black flammules; aperture proportionally wide, becoming wider toward anterior end; interior of aperture dark brownish-purple with 2 wide white bands.

Type Species: *Conus hieroglyphus* Duclos, 1833, endemic to Aruba Island, Dutch West Indies (ABC Islands), by monotypy.

Etymology: Named as a combination of Aruba and *Conus*, in reference to the new genus being endemic to the island of Aruba.

Discussion: The only known species of *Arubaconus*, *A. hieroglyphus*, resembles no other living western Atlantic cone shell. With its stumpy inflated body whorl, rounded shoulder, and turbinate spire, the living Aruban endemic species somewhat resembles the fossil *Conus isomitratus* Dall, 1896 from the Burdigalian Miocene Chipola formation of northern Florida. This early Miocene species differs from the living *A. hieroglyphus* in being a larger shell and in lacking the strong beaded spiral cords around the body whorl. The Florida fossil may represent the oldest known member of *Arubaconus*, demonstrating that the genus evolved in the Chipolan Subprovince of the Baitoan province. If so, then *A. hieroglyphus* can be considered a Miocene relict taxon.

Attenuiconus new genus

Diagnosis: Shells of small to average size for subfamily, very elongated, with straight sides and narrow, straight apertures; spires low or flattened, with projecting, mamillate protoconchs of 2 or 3 whorls; spire whorls may be flattened or slightly caniculate; shells generally smooth and polished, but some species have coarse sculpture of fine spiral threads; shells generally colored in yellows or oranges arranged in wide bands, but may also be colored pink, salmon, or reddish-orange with brown or white longitudinal flammules.

Type Species: *Conus attenuatus* Reeve, 1844, tropical western Atlantic from southern Florida to the Lesser Antilles and Barbados.

Other Species in *Attenuiconus*: *A. eversoni* (Petuch, 1987), Roatan Island; *A. honkeri* (Petuch, 1988), Venezuela; *A. ignotus* (Cargile, 1998), central Caribbean Banks; *A. poulosi* (Petuch, 1998), Venezuela and northern Colombia.

Etymology: Named as a combination of *attenuatus* (Latin, "stretched out" or "made thin") and *Conus*, in reference to the characteristic elongated shape of this small group of cone shells.

Discussion: Members of this new genus were previously referred to *Dauciconus* Cotton, 1945, but differ from members of that genus by having much more elongated and narrow shells and in having the projecting, needle-like protoconch. The genus is confined to the southern part of the Carolinian province and the entire Caribbean province. *Attenuiconus attenuatus* (= *ustickei*) is the widest-ranging species in its genus, being found near the deep reefs off Palm Beach and Broward Counties, southeastern Florida, throughout the Caribbean Basin, and throughout the Lesser Antilles to Barbados. Two of the *Attenuiconus* species, *A. honkeri* and *A. poulosi*, are found only off the coasts of northern Colombia and the Gulf of Venezuela area, where they occur on deep-water sponge reefs.

Bermudaconus new genus

Diagnosis: Shells elongated, fusiform, with high, pyramidal spires and completely rounded shoulders, which are only slightly angled; shells smooth and polished, with silky texture; anterior tip encircled by numerous very fine spiral threads; color patterns highly ornate, generally composed of bright pinkish-orange base color overlaid by 2 wide bands of alternating dark brown and white flammules, one around the midbody and one between midbody and shoulder; anterior end with white and brown flammules; thin band of alternating brown and white dashes and dots around area between midbody and anterior end; spire white, marked with large, amorphous brown flammules; aperture narrow, colored pink or pinkish-white on inside.

Type Species: *Conus lightbourni* Petuch, 1986, by monotypy, from deep-water areas around the Bermuda Seamount.

Etymology: Named as a combination of Bermuda and *Conus*, reflecting the endemic nature of the genus.

Discussion: Only a single species, *Bermudaconus lightbourni*, is known to belong to this new genus. With its rounded shoulder area that blends directly into the spire whorls and with its bright, almost gaudy, colors, the sole member of this genus resembles no other western Atlantic cone shell. The genus appears to be endemic to the Bermuda Islands and Seamount and evolved as a result of its extreme isolation from the rest of the tropical western Atlantic.

Brasiliconus new genus

Diagnosis: Shell small, fusiform, biconic, with high elevated spires and straight sides; shoulder sharply angled, ornamented with numerous large rounded knobs; spire whorls coronated, ornamented with small rounded knobs; body whorl smooth, sculptured with very faint, fine spiral threads; aperture narrow, straight; shell color white, overlaid with variable amounts of small brown flammules and rows of brown dots and dashes.

Type Species: *Conus scopolorum* Van Mol, Tursch, and Kempf, 1971, by monotypy. From offshore northern Brazil and the Islands of Fernando de Noronha and Atol das Rocas.

Etymology: Named as a combination of Brazil and *Conus*, reflecting the endemic nature of the genus.

Discussion: Although resembling members of the Caribbean endemic genus *Tenorioconus*, the new genus *Brasiliconus* differs in consistently having a smaller shell with a proportionally higher spire, far more protracted spire. Unlike *Tenorioconus*, the new Brazilian genus is distinctly biconic, often with a spire that is equal in length to the body whorl. *Brasiliconus*, represented by a single species, is confined to the Cearaian Subprovince of the Brazilian Molluscan Province and ranges from Amapa State to Fernando de Noronha Island and Canopus Bank.

Genus *Dauciconus* Cotton, 1945

Dauciconus jorioi new species
(Figure 10.14A,B)

Description: Shell of average size for genus, heavy and thickened, broadly conical, with straight sides; spire slightly elevated, broadly subpyramidal, distinctly stepped; shoulder sharply angled, edged by thin carina; body whorl smooth and silky (slightly pitted and eroded on holotype), with 3 faintly raised spiral cords around anterior tip; shell base color white, overlaid with 2 broad bands of reddish-brown and orange-brown amorphous flammules, one around posterior two thirds of shell and one around anterior one third of shell; broad brown bands separated by wide white band edged by dark brown, widely separated tooth-shaped flammules; edge of shoulder and posterior one third of shell marked with intermittent, evenly separated large white longitudinal flammules; spire white with scattered small dark reddish-brown crescent-shaped flammules; anterior tip of shell salmon-orange; aperture proportionally very narrow, pale salmon on interior.

Holotype: Length 53 mm, width 29 mm, FMNH 328432.

Type Locality: On open rock platform bottom, occupied by a hermit crab, in 5 m depth off the southern coast of Trindade Island, Espiritu Santo State, Brazil.

Etymology: Named for Domingos Afonso Jorio of Guaraparí, Espiritu Santo State, Brazil, in honor of his many contributions to Brazilian malacology.

Discussion: This new *Dauciconus* species is the largest cone shell found on Trindade Island, part of the isolated and remote Trindade and Martim Vaz Islands, which occur over 800 km due east of Vitoria, Espiritu Santo State, Brazil. Of the known Brazilian *Dauciconus* species, *D. jorioi* is most similar to *D. riosi* (Petuch, 1986) from the coastal areas off Bahia State but differs in being a smaller, stockier, less elongated shell that is proportionally broader across the shoulder and in having a higher, more elevated, and broadly pyramidal stepped spire. Although sharing the same type of color pattern with the closely related Brazilian coastal species *D. riosi* and *D. worki* (Petuch, 1998), the new Trindade Island endemic differs in shell shape and proportions and represents yet another member of the interesting Brazilian *Dauciconus* species complex.

Kellyconus new genus

Diagnosis: Shells generally small for family, stocky, broad across shoulders, with low, subpyramidal spires; shoulders angled, smooth, often bounded by low, rounded carinae; body whorls usually heavily sculptured with coarse spiral cords and threads; body whorls sometimes ornamented with low, longitudinal undulations and wrinkles, producing distinct corrugated appearance; spire whorls sculptured with coarse spiral cords; apertures narrow; shells usually colored in bright hues, usually reds, oranges, yellows, pinks, or violets, or combinations of all of these, although some species are colored white with yellow-orange patches or pale tan; color patterns arranged as series of large patches in spiral rows or rows of small dots.

Type Species: *Conus patae* Abbott, 1971, from the Caribbean region, ranging from southeastern Florida to the Lesser Antilles.

Other Species in *Kellyconus*: *K. binghamae* (Petuch, 1987), off southeastern Florida; *K. rachelae* (Petuch, 1988), off Venezuela and Suriname.

Etymology: Named for Kelly McCarthy, wife of Andre Poremski, an inspired amateur naturalist and diver.

Discussion: This new genus of small, highly colored shells is found primarily on living coral reefs throughout the Caribbean region. Here, *Kellyconus* species live buried deeply within coral rubble and are generally hard to collect. Although resembling species of *Dauciconus*, members of this new genus have consistently smaller, more truncated, and more sculptured shells. The two genera, *Dauciconus* and *Kellyconus*, are most likely sister groups, having evolved from a common ancestor in the Miocene or Pliocene.

Genus *Lindaconus* Petuch, 2002

Lindaconus therriaulti new species
(Figure 5.6F)

Description: Shell smaller than typical *Lindaconus spurius*, stocky, broad across shoulder, distinctly pyriform; shoulder angled, bordered by large rounded carina; spire high and protracted, with spire whorls slightly canaliculate; shell smooth and silky, with numerous very low, fine threadlike cords around anterior end; base color pure white, overlaid with 12–15 rows of small, rounded dark purple-brown, orange-brown (as on holotype), or yellow-orange dots; some specimens pure white, without spots; spire whorls white with large, evenly-spaced dark brown rounded or oval spots; aperture narrow, white on interior.

Holotype: Length 43 mm, width 25 mm, FMNH 328433.

Type Locality: Trawled by commercial shrimpers from 35 m depth on Campeche Bank, off Progresso, Yucatan State, Mexico.

Etymology: Named for Lyle Therriault of Concord, North Carolina, inspired amateur naturalist, artist, and lover of the Conidae.

Discussion: This new Yucatanean Subprovince species is most similar to the widespread Carolinian species *Lindaconus atlanticus* (Clench, 1942), from the Floridian, Suwannean, and Texan Subprovinces (which was originally described as a subspecies of the Caribbean province *L. spurius*, but has now been shown to be a distinct, valid species that is confined to the Carolinian province). *Lindaconus therriaulti* differs from *L. atlanticus* in being consistently a smaller, stockier shell that is characteristically pear-shaped and in having a color pattern composed of small, widely separated brown dots that never connect together or form large amorphous patches as on typical *atlanticus*. By having a distinct pyriform shell shape and a color pattern of small brown dots, *Lindaconus therriaulti* often has been confused with the sympatric but unrelated *Gradiconus sennottorum* (Rehder and Abbott, 1951). Pure white specimens of the new pear-shaped Yucatan species are occasionally collected, and these have often been confused with the larger, more straight-sided and more elongated *L. aureofasciatus* (Rehder and Abbott, 1951), which occurs in deep water along the Suwannean Subprovince.

Genus *Magelliconus* da Motta, 1991

Magelliconus eleutheraensis new species
(Figure 6.8G)

Description: Shell of average size for genus, elongated, straight-sided; spire low, subpyramidal; shoulder proportionally broad, heavily ornamented with 11–12 large rounded, prominent knobs; body whorl shiny and polished, ornamented with 8–10 large, widely spaced, heavy beaded spiral cords; shell colored bright orange-red or red (in fresh live specimens), with scattered large white flammules and blotches; most shoulder knobs white, separated by red areas; spire white, with scattered amorphous large red and orange flammules; protoconch and early whorls bright red; aperture very narrow, with violet interior; anterior tip pale purple, encircled with 8 very fine spiral threads.

Holotype: Length 19 mm, width 11 mm, FMNH 328434.

Type Locality: Found among coral and shell rubble on the beach at the base of the eroded cut (the "Windows") between the northern and southern part of Eleuthera Island, Great Bahama Bank, Bahamas.

Etymology: Named for Eleuthera Island, Bahamas, to which the species is endemic.

Discussion: Of the known Bahamian *Magelliconus* species, the new Eleuthera taxon is closest to *Magelliconus jacarusoi* (Petuch, 1998) from Paradise Island and the New Providence Island area. The new species differs from the similarly colored *M. jacarusoi* in having a more elevated spire, in having fewer and proportionally much larger shoulder knobs, and in having the prominent, large beaded spiral cords on the body whorl. The holotype (and all other specimens examined by the senior author) was collected dead on the beach on the western side of the narrow, gulley-like cut across Eleuthera, known locally as "The Windows." This area typically has high surf and strong, dangerous currents, making diving and collection conditions virtually impossible. The new cone apparently lives in these high surf areas, probably in shallow water, and for that reason has only been collected as dead, often heavily eroded, beach specimens. Live individuals are probably a deep red with white flammules.

Poremskiconus new genus

Diagnosis: Shells small for family, short, stocky, broad across shoulders; spires low, broadly pyramidal, flattened on many species; shoulders sharply angled, frequently bordered with low, rounded carina, producing slightly canaliculate whorls; body whorls smooth and shiny, frequently sculpted with very faint, low spiral threads; spire whorls with 2 or 3 low spiral cords; apertures narrow; shells generally brightly colored, bearing hues of pink, red, orange, khaki green, blue, tan, or black, often with combinations of several of these colors; color patterns generally composed of large amorphous patches or wide bands, often overlaid with rows of dots; spire whorls often colored with large crescent-shaped flammules.

Type Species: *Conus archetypus* (Crosse, 1865), coast of Bahia State, Brazil.

Other Species in *Poremskiconus*: *P. abrolhosensis* (Petuch, 1987) (= *baiano* Coltro, 2004), Abrolhos Archipelago, southern Bahia State, Brazil; *P. archetypus* (Crosse, 1865), Bahia State, Brazil; *P. colombi* (Monnier and Lampalaer, 2012), Martinique; *P. beddomei* (Sowerby, 1901), Grenadines, Lesser Antilles; *P. bertarollae* (Costa and Simone, 1997), Bahia State, Brazil; *P. brasiliensis* (Clench, 1942), Espititu Santo coast, Brazil; *P. cargilei* (Coltro, 2004), Bahia State, Brazil; *P. colombianus* (Petuch, 1987), Rosarios Islands, Colombia; *P. edward-puuli* (Petuch, 1998), San Blas Islands, Panamá; *P. hennequini* (Petuch, 1992), Martinique and Guadalupe Islands, Lesser Antilles; *P. mauricioi* (Coltro, 2004), Bahia State, Brazil.

Etymology: Named for Andre Poremski, master diver, underwater photographer, and one of the consummate experts on the western Atlantic cone shells.

Discussion: This compact group of small, smooth-shouldered cones had previously been considered to belong with either the coronated *Magelliconus* species complex or with the tiny *Curiboconus* species. Having noncoronated, smooth shoulders, the members of *Poremskiconus* differ greatly from the similar-shaped but distinctly knobbed-shouldered species of *Magelliconus*. Although having smooth shoulders like *Cariboconus* species, members of *Poremskiconus* differ in having larger shells (more than twice as large on average) with proportionally higher and more pyramidal spires and in being wider and broader in outline. Although essentially a Brazilian genus (being more abundant and better represented in the Brazilian Molluscan province), *Poremskiconus* does range within the southern Caribbean as far north as Martinique and the Grenadines in the Lesser Antilles and the San Blas Islands of Panamá.

Sandericonus new genus

Diagnosis: Shells of average-to-small size for family, generally very elongated, with straight sides and flattened spires; body whorls smooth and silky-textured, frequently sculptured with very numerous extremely fine spiral threads; shoulders sharply angled, bordered by low, rounded carina; spire whorls sculptured with fine spiral threads; apertures narrow, straight; shells generally brightly colored, usually with cloudings of bright orange, pink, violet, or blue, often overlaid with large flammules of brown or tan.

Type Species: *Conus carioca* Petuch, 1986, from offshore of southern Espiritu Santo and Rio de Janeiro States, Brazil.

Other Species in *Sandericonus*: *S. ednae* n. sp., Brazil; *S. hunti* (Wils and Moolenbeek, 1979), Barbados; *S. knudseni* (Sander, 1982), Barbados; *S. perprotractus* (Petuch, 1987), Venezuela, Barbados, Guyana, and Suriname; *S. sanderi* (Wils and Moolenbeek, 1979), Barbados; *S. sorenseni* (Sander, 1982), Barbados and Guyana.

Etymology: Named in honor of Dr. Finn Sander, professor of biology and former director of the McGill University Marine Laboratory at St. James, Barbados, in recognition of his many contributions to our knowledge of the deep-water ecology of the Barbados Seamount.

Discussion: With their straight-sided shells and flat spires, members of *Sandericonus* somewhat resemble slender specimens of *Dauciconus*. Species in this new genus differ in having more elongated shells with proportionally flatter spires; in having a silky shell sculpture produced by numerous fine spiral threads; and in having color patterns made up of amorphous clouding and distinct rows of large patches and rectangular dots. *Sandericonus* species also live in much deeper water than do *Dauciconus* species, preferring depths of anywhere from 50 m to over 400 m, and are confined to offshore ecosystems. Around the isolated Barbados Seamount, a complex of at least four *Sandericonus* species has evolved in bathyal depths, where they occur with the deep-water conid and conilithid genera *Conasprelloides* and *Dalliconus*. The new genus is confined to the extreme southern Caribbean Province (Barbados and the Surinamian Subprovince) and the Brazilian Province.

Sandericonus ednae new species
(Figure 11.6C,D)

Description: Shell of average size for genus, shiny and polished, slightly truncated in outline, with proportionally wide shoulder; early whorls of spire excerted and scalariform, with later whorls becoming distinctly flattened; shoulder sharply angle, edged with large rounded carina, producing distinctly canaliculate spire whorls; spire whorls ornamented with 3 thin spiral threads; anterior third of body whorl ornamented with numerous very low, faint spiral cords; shell color pale pinkish-white with 3 wide bands of pale orange amorphous flammules, one around edge of shoulder, one around midbody, and one around anterior third; 4 widely separated rows of pale orange-tan dots present on body whorl, with one row along anterior side of orange shoulder band, 2 rows on either side of midbody band, and one row along posterior side of anterior band; spire pinkish-white, with numerous evenly spaced orange crescent-shaped flammules; aperture uniformly narrow; interior of aperture white; protoconch proportionally large, rounded, composed of 2 whorls, tan in color.

Holotype: Length 29 mm, width 15 mm, FMNH 328431.

Type Locality: Trawled by commercial shrimp boat from 50 m depth off Farol de São Tome, Rio de Janeiro State, Brazil.

Etymology: Named for Edna Aguilar Jorio, inspired amateur naturalist and wife of Brazilian malacologist Afonso Jorio.

Discussion: This new Brazilian *Sandericonus* species is closest to the congeneric *S. carioca* (Petuch, 1986), also from offshore areas of Rio de Janeiro State, but differs in being a smaller shell with a stumpier, less elongated, and more pyriform shape; in being a less colorful shell; and in having pale pastel orange bands instead of the bright orange, brown, tan, and pink bands seen on *S. carioca*. The spires of the two closely-related species are also very different, with the spire whorls of *S. ednae* being slightly canaliculate and ornamented with three faint spiral threads and with those of *S. carioca* being much flatter, lacking the large shoulder carina, and being ornamented with only one or two faint threads near the suture. In size and general shape, the new species is also similar to *Sandericonus sanderi* (Wils and Moolenbeek, 1979) from deep water off Barbados but differs in having a paler shell color; in having slightly canaliculate spire whorls; and in having less sculptured spire whorls that lack the five strong spiral cords that are seen on *S. sanderi*. *Sandericonus ednae* is part of a highly endemic deep-water fauna that is found off Cabo de São Tome, an area of faunal overlap between the extreme southernmost edge of the Bahian Subprovince and northernmost edge of the Paulinian Province.

Tuckericonus new genus

Diagnosis: Shells small for family, elongated, fusiform, with straight sides; spires of variable height, most often elevated and subpyramidal; shoulders sharply angled, bordered by low, rounded carina; body whorls smooth with silky texture, often faintly sculptured with very fine spiral threads and faint cords; anterior ends of shells encircled by numerous deeply incised spiral grooves; spire whorls smooth, sometimes with few very faint spiral threads; apertures narrow and straight; protoconchs characteristically very pronounced, projecting, needle-like, forming elongated mamillate structure composed of 2 or 3 whorls; shells mostly colored in shades of bright yellow, orange, pink, brown, or red, frequently overlaid with large brown or tan flammules and thin spiral lines, often with large white patches.

Type Species: *Conus flavescens* Sowerby, 1834, from shallow carbonate sand areas near coral reefs along southeastern Florida, the Florida Keys and northern Cuba, the Bahamas, and south to the Turks and Caicos.

Other Species in *Tuckericonus*: *T. flamingo* (Petuch, 1980), 60 m depths off southeastern Florida and the Florida Keys; *T. ceruttii* (Cargile, 1998), Bay Islands of Honduras and the Central Caribbean Banks; *T.* unnamed species, with a multinucleate protoconch, off southeastern Florida.

Etymology: Named for John Tucker of Grafton, Illinois, one of the leading experts on the higher systematics of the Conoidean gastropods, in recognition of his many contributions to our knowledge of the western Atlantic cone shells.

Discussion: Members of this small but distinctive group of cone shells have traditionally been placed in the genus *Dauciconus*, primarily because of their general shell shape and colors. *Tuckericonus* differs from *Dauciconus*, however, in having consistently smaller, more elongated, and narrower shells, with more projecting early whorls and protoconchs. Members of the new genus lack the heavy spiral cords frequently seen on the spires of *Dauciconus* species and have more deeply incised spiral grooves around the anterior end. The protoconch structure of *Tuckericonus* is the most outstanding feature of the genus, being characterized by having unique needle-like projecting whorls that extend well beyond the early whorls of the teleoconch. The genus is confined to the southern end of the Carolinian Molluscan province (Floridian Subprovince) and the Bahamian and Nicaraguan Subprovinces of the Caribbean Molluscan province. The Bahamian and Floridian specimens of *T. flavescens* may prove to represent two distinct species. If this is the case, then true *T. flavescens* would be confined to the Bahamas and the Turks and Caicos and the Floridian species would be referable to *T. caribbaeus* (Clench, 1942) (for specimens named from off Palm Beach County).

Family Conilithidae
Subfamily Conilithinae
Genus *Dalliconus* Tucker and Tenorio, 2009

Dalliconus coletteae new species
(Figure 9.6F,G)

Description: Shell small for genus, averaging around 22 mm, narrow, slender, with slightly convex sides; shoulder sharply-angled, ornamented with 20–24 small rounded beadlike knobs per whorl; subsutural area sharply sloping; spire highly elevated, protracted, scalariform, approximately one-third length of entire shell; body whorl ornamented with 30–32 deeply-incised thin spiral sulci, producing grooved, rough-textured appearance; body whorl color pale cream-white to white, overlaid with 4 bands of light brown rectangular spots and scattered large longitudinal amorphous flammules; in some specimens, rectangular spot pattern dominates while in others (like the holotype) longitudinal flammules dominate; spire whorls with 12–14 evenly spaced small crescent-shaped flammules per whorl; protoconch white, proportionally large and bulbous, composed of two and one-half whorls; aperture uniformly narrow, straight, white within interior.

Holotype: Length 20 mm, width 7 mm, FMNH 328436; **Paratype:** length 21 mm, same locality and depth as holotype, in the collection of Jonnie Kuiper, Veendam, Netherlands; another specimen, length 21 mm, from the same locality as the holotype, in the research collection of the author.

Type Locality: Dredged from 300 m depth off St. James, Barbados.

Etymology: Named for Colette Kuiper-Hoorn of Veendam, Netherlands.

Discussion: Of the known *Dalliconus* species of the Antillean and Grenadian Subprovinces, *D. coletteae* comes closest to *D. roberti* (Richard, 2009) from deep water off Guadeloupe. The new Barbados species differs from *D. roberti* in being a much smaller and less elongated shell, in having a more sculptured body whorl with numerous deeply-incised spiral sulci, and in being a less colorful shell, having only scattered small flammules and spots and not the densely-packed pattern seen on the Guadeloupe congener. Of the known western Atlantic members of this species complex, *D. coletteae* is most similar to *D. macgintyi* (Pilsbry, 1955) from the Carolinian Province, but differs in being a much smaller, less elongated, and more coarsely grooved shell that lacks the densely packed brown patches and longitudinal flammules typically seen in *D. macgintyi*. At present, *Dalliconus colletteae* has been collected only off Barbados, but may extend farther along the Lesser Antilles and ABC Islands.

Coltroconus new genus

Diagnosis: Shells very small for family and subfamily, averaging only 12 mm in length, broadly conical, proportionally wide across shoulder, with straight sides; body whorls generally smooth and polished, with some species having subdued ornamentation composed of low, beaded cords or slightly impressed spiral grooves; shoulders sharply angled, typically ornamented with large, prominent, rounded knobs; spires proportionally low, flattened, subpyramidal, with low, undulating knobs ornamenting spire whorls; shell colors generally white or pinkish-white with variable amounts of large red or brown patches, sometimes overlaid with rows of small brown dots; some species being predominantly solid red or brown in color; apertures proportionally wide, becoming wider toward anterior tip; protoconchs proportionally large, mammilate, rounded, composed of 2 whorls.

Type Species: *Conus iansa* Petuch, 1979, Abrolhos Archipelago, Bahia State, Brazil (Figure 10.13A,B).

Other Species in *Coltroconus*: *Coltroconus delucai* (Coltro, 2004), on reef complexes, southern Bahia State, Brazil (Figure 10.7B,C; *C. bodarti* Coltro, 2004, is a synonym); *Coltroconus schirrmeisteri* Coltro, 2004, seamounts and deep reefs off southern Bahia State and northernmost Espiritu Santo State, Brazil. *Coltroeonus bodarti* (Coltro, 2004) is now known to be a synonym of *C. iansa*.

Etymology: Named for Jose and Marcus Coltro of São Paulo, Brazil, in recognition of their many discoveries and contributions to Brazilian malacology.

Discussion: This group of tiny cones, containing the smallest species of western Atlantic Conilithidae, is confined to southern Bahia State, the Abrolhos Archipelago area, and the seamounts off northernmost Espiritu Santo State, Brazil. The new genus most probably represents an endemic species radiation that is centered on the Abrolhos Archipelago and reef complexes, where it is most commonly encountered. Of the known conoliLthids, *Coltroconus* species most closely resemble members of the Brazilian *Jaspidiconus* species complex but differ in consistently having much smaller, stumpier, and more truncated shells that are proportionally wider across the shoulder; in having lower, flatter spires with sharply angled shoulders; in having prominent large rounded shoulder knobs; and in having a straight-sided profile instead of a convex one as is typically seen in *Jaspidiconus*.

Genus *Jaspidiconus* Petuch, 2004

Jaspidiconus allamandi new species
(Figure 7.11F)

Description: Shell of average size for genus, elongated and cylindrical, with slightly convex sides; shoulder angled, edged with strong rounded cord; substural area sharply sloping to edge of shoulder; spire high, pyramidal, with slightly convex whorls; shell sculptured with 15–16 large rounded spiral cords, each pair separated by deep spiral groove, giving shell rough appearance; spiral cords on anterior half of shell ornamented with small beads; spire whorls and subsutural area with single low, broad cord; aperture narrow, flaring slightly at anterior end; shell color deep chocolate-brown or blackish-brown, overlaid with variable amounts of pale blue or bluish-white spotting and small amorphous flammules; spiral cords often with alternating pale blue and chocolate-brown spots; interior of aperture dark purple-brown.

Holotype: Length 16.2 mm, width 7.2 mm, FMNH 328435; **Paratypes:** lengths 16.3 mm, 17.7 mm. and 18.5 mm, same locality as the holotype, in the research collection of Andre

Poremski, Washington, D.C.; length 16.7 mm, same locality as the holotype, in the research collection of the author.

Type Locality: On silty sand in turtle grass (*Thalassia*), 3 m depth in Sandy Bay, northwestern coast of Roatan Island, Bay Islands, Honduras.

Etymology: Named for Randy Allamand of Sebring, Florida, renowned diver and commercial shell dealer, who collected the type material on Roatan Island.

Discussion: In size and basic type of ribbed sculpture pattern, *Jaspidiconus allamandi* most closely resembles the Roatan Island endemic species *J. roatanensis* Petuch and Sargent, 2011. The new species differs in being a more cylindrical, less inflated shell; in being much more darkly colored; and in having a dark brown shell color with bluish-white spots as opposed to a white shell color with reddish-brown zebra stripes (see Petuch and Sargent, 2011, for a color series of *J. roatanensis*). With its dark shell color and rows of blue-white spots, *J. allamandi* also resembles the Antillean (Puerto Rico to Grenada) *J. duvali* (Bernardi, 1862) (= *J. boubeeae*; Sowerby, 1903) but differs in being a much more coarsely sculptured shell and in having numerous rows of cords and deeply impressed spiral sulci as opposed to a smooth shell as seen in *J. duvali*. As far as is known, *J. allamandi* is endemic to the turtle grass beds of northern Roatan Island.

Jaspidiconus exumaensis new species
(Figure 6.8C,D)

Description: Shell small for genus, elongated, with high pyramidal stepped spire; shoulder sharply angledcarinated; body whorl with distinct indentation around midbody, producing characteristic convex outline; body whorl shiny and polished, sculptured with 10 incised spiral grooves around anterior two thirds of body whorl; anterior tip encircled with 3–5 small spiral threads; shell color white or cream-white with 8–10 evenly spaced tiny brown dots along edge of shoulder carina and spire whorls; aperture narrow, with interior being white or cream-white in color; protoconch proportionally large, exserted, mamillate, composed of 2.5 whorls.

Type Material: Holotype, length 15.2 mm, width 7 mm, FMNH; **Paratypes:** length 14 mm and length 12.5 mm, both same locality as the holotype, in the research collection of the author.

Type Locality: In sand, 2 m depth off Cape Eleuthera, southwestern side of Eleuthera Island, Bahamas, along the southern Exuma Sound.

Etymology: Named for the Exuma Sound of the Bahamas.

Discussion: This new Great Bahama Bank endemic cone is closest to *Jaspidiconus branhamae* (Clench, 1953) from the Abaco Islands, Little Bahama Bank, but differs in being a much smaller shell with distinctly indented sides and a concave profile; in being a less colorful shell that lacks any large brown color patches and orange-brown flammules; and in having the widely spaced incised spiral grooves around the body whorl. In the shallow water sand areas of Cape Eleuthera Bay on the Exuma Sound, *Jaspidiconus exumaensis* occurs together with the much larger, heavily pustulated *J. nodiferus* (Kiener, 1845) and the bright yellow *Tuckericonus flavescens* (Sowerby, 1834) but is never as common as these two species. The Bahamas Islands are now known to house at least six distinct species of *Jaspidiconus*; *J. verrucosus* (Hwass, 1792) (Little Bahama Bank and western Great Bahama Bank), *J. nodiferus* (eastern Great Bahama Bank and Turks and Caicos), *J. exumaensis* (Exuma Sound area), *J. mindanus* (Hwass, 1792) (Little and Great Bahama Banks and Turks and Caicos), *J. branhamae* (Clench, 1953) (Abaco Islands, Little Bahama Bank), and *J. oleiniki* new species (described later in this section) (Bimini Chain, Great Bahama Bank).

Jaspidiconus mackintoshi new species
(Figure 6.15D,E)

Description: Shell of average size for genus, elongated, with straight sides; shoulder sharply angled, bordered by sharp carina; spire pyramidal, stepped; body whorl smooth and shiny, with numerous very fine, faintly incised spiral sulci which become larger and more prominent at anterior end; aperture narrow, becoming slightly wider at anterior end; shell color pale salmon-pink with variable amounts of amorphous orange flammules; orange and pink base color overlaid with 20–30 very fine, closely packed reddish-brown spiral hairlines and scattered small white patches; shoulder carina white, marked with line of widely spaced large orange spots; spire whorls pale whitish-salmon with evenly spaced large pale orange flammules that correspond to carina spots; early whorls and protoconch white; interior of aperture pale salmon-pink; protoconch proportionally very large, mamillate, projecting, composed of 2 whorls.

Holotype: Length 13.7 mm, width 7.1 mm, FMNH 328439; **Paratypes:** lengths 13.2 mm, 13.6 mm, 14.4 mm, and 15.7 mm, same locality as the holotype, in the research collection of Andre Poremski, Washington, D.C.; length 13.4 mm, same locality as the holotype, in the research collection of the author.

Type Locality: In clean carbonate sand and coral rubble, 10 m depth off Little St. James Island, U.S. Virgin Islands.

Etymology: Named for Gary Mackintosh, renowned Caribbean collector and diver, who discovered the new species in the Virgin Islands.

Discussion: With its distinctive pattern of revolving fine brown spiral hairlines, *Jaspidiconus mackintoshi* resembles no other known western Atlantic *Jaspidiconus* species. At first glance, the new Virgin Islands cone could be confused with dwarf specimens of *Jaspidiconus mindanus* form *karinae* (Nowell-Usticke, 1968), which is found on St. Croix, but differs in being a much smaller, narrower, and more elongated shell; in having a proportionally much larger, more projecting protoconch; and in having a color pattern composed of closely packed fine brown spiral lines. *Jaspidiconus mackintoshi* appears to be endemic to the U.S. Virgin Islands.

Jaspidiconus oleiniki new species
(Figure 6.8E,F)

Description: Shell small for genus, broad across shoulder; sides straight, tapering to anterior end; shoulder sharply angled, bordered by low rounded carina; spire whorls elevated, slightly concave and canaliculate, wide and distinctly pyramidal; body whorl smooth and shiny, ornamented with 10–12 deeply incised spiral grooves around anterior one third of shell; shell color variable, ranging from pure white (like paratype) to pale cream-white, overlaid with large amorphous pale orange flammules (like holotype); edge of shoulder carina marked with large, widely spaced orange dots; spire whorls with small amorphous pale orange flammules; aperture straight, narrow, white or pale orange on interior; protoconch proportionally large, rounded, bulbous, composed of 2 whorls.

Type Material: Holotype, length 15 mm, width 8.5 mm, FMNH 328440; Paratypes; length 13 mm, FMNH, same locality as holotype; length 16 mm, same locality as holotype, in the research collection of the author; 10 specimens, lengths 9-20.6 mm, in the research collection of Anton Oleinik.

Type Locality: In clean oölitic sand, 1 m depth in Nixon's Harbour, South Bimini Island, Bimini Chain, Great Bahama Bank, Bahamas.

Etymology: Named for Dr. Anton Oleinik, Department of Geosciences at Florida Atlantic University, who collected the type material on South Bimini Island.

Discussion: Of all the known Bahamian cone shells, this new species most closely resembles the previously described *Jaspidiconus exumaensis*, especially in its small size. The new Bimini *Jaspidiconus*, however, differs from *J. exumaensis* in being a stockier shell with a proportionally wider shoulder; in having a wider, more pyramidal spire; in being a smoother shell, with incised grooves only around the anterior one-third of the body whorl; and in having a much straighter shell profile, lacking the distinct indented "waist" (concave sides) that is so indicative of the Exuma Sound species. *Jaspidiconus oleiniki* is also a much more colorful shell than the pure white or pale cream-colored *J. exumaensis*, often being tinted with large, pale orange flammules. At present, *J. oleiniki* is known only from the Bimini Chain of islands, along the western edge of the Great Bahama Bank.

Jaspidiconus sargenti new species
(Figure 7.6G)

Description: Shell of average size for genus, shiny and polished, subfusiform, with rounded convex sides; shoulder sharply angled, edged with large, prominent rounded carina; spire elevated, subpyramidal, distinctly stepped; subsutural area and spire whorls depressed, shallowly canaliculate; posterior half of body whorl smooth and shiny, with few faint inscribed spiral threads; anterior half of body whorl heavily sculptured with 10 prominent, deeply-incised spiral sulci, becoming deeper and more prominent toward anterior tip; shell uniformly pale cream-white with few widely-scattered very pale yellow amorphous flammules on body whorl; carinae of spire whorls sometimes marked with widely-separated tiny pale tan dots; protoconch proportionally large, composed of two and one-half whorls, pale yellow-tan in color; early whorls pale yellow-tan; aperture uniformly narrow, widening slightly at anterior end; interior of aperture cream-white in color.

Holotype: Length 21.7 mm, width 10.75 mm, FMNH 328441; an additional specimen, length 20.2 mm, from the same locality as the holotype, in the research collection of the author.

Type Locality: Trawled by commercial shrimpers from 30 m depth, 20 km southeast of Roatan Island, Bay Islands, Honduras.

Etymology: Named for Dennis M. Sargent of Mount Dora, Florida, inspired amateur malacologist, naturalist, and photographer, in recognition of his many contributions to the systematics of the Olividae, Strombidae, and Conidae.

Discussion: Of the known *Jaspidiconus* species of the Central Caribbean Basin, *J. sargenti* is most similar to *J. roatanensis* Petuch and Sargent, 2011 (Figure 7.7A, Chapter 7) from Roatan Island, but differs in being a larger, thinner, and more inflated shell; in having a smoother, less sculptured shell that lacks strong beaded cords; and in being a much less colorful shell, having an unmarked, mostly white shell that lacks the dark zebra-pattern flammules that are seen on the Roatan endemic. Recently, commercial shell dealers and shell collectors have incorrectly referred to this new species as "*Conus commodus* A. Adams, 1855", a *nomen dubium* that most probably is a synonym of the conid *Gradiconus philippii* (Kiener, 1845) or some other closely-related *Gradiconus* species.

Bivalvia

Heterodonta
Veneroidea
Family Veneridae
Subfamily Chioninae
Genus *Mercenaria* Schumacher, 1817

Mercenaria hartae new species
(Figure 3.6E)

Description: Shell small for genus, thin, ovately rounded in outline; lunule proportionally large, heart-shaped, bounded by incised line; inner margin of commissure crenulate; umbones recurved, strongly prosogyrate; shell exterior sculptured with numerous strong, rough, raised concentric growth lines and large, widely separated concentric lamellae; prominent large, smooth, and shiny semicircular patch present in area ranging from edge of commissure to center of valve; shiny semicircular patch bounded anteriorly and posteriorly by strong raised lamellae; shell color pale yellow-cream, with pale purple amorphous patches and bands present on escutcheon and near umbones; pallial sinus deeply reflexed and sharply pointed; interior of shell pale yellow-cream, often marked with band of pale purple along posterior edge of commissure; both valves with 3 cardinal teeth, with central tooth bifurcated by deep groove.

Holotype: Width 44.28 mm, height 39.51 mm, FMNH 328438; additional specimens, width 40 mm, same locality as holotype, in the research collection of the author; width 42 mm, in the Jane Hart collection.

Type Locality: On intertidal sandy mud flats near red mangrove forests along Pine Point, Singer Island, Lake Worth Lagoon, Palm Beach County, Florida.

Etymology: Named for Jane Hart, inspired amateur naturalist and renowned secondary school science educator (now retired from the Palm Beach County School Board), in recognition of her lifetime of contributions to science education in Florida.

Discussion: This previously overlooked *Mercenaria* species is confined to the area of the Palm Beach Provinciatone (from Fort Pierce south to Lake Worth), where it is an inhabitant of muddy coastal lagoons and inland tidal creeks near red mangrove forests. Throughout its distribution, *Mercenaria hartae* is sympatric with the wide-ranging *Mercenaria campechiensis* (Gmelin, 1791) but can be easily separated by several consistent differences in shell morphology: (1) *M. hartae* is a much smaller and flatter species with a thinner shell; (2) *M. campechiensis* is heavily sculptured with fine concentric ribs and lamellae over its entire shell surface, while *M. hartae* has a very prominent and distinctive shiny and unsculptured semicircular patch that extends from the commissure and edge of the shell to the center of the valve; (3) the shell of *M. hartae* is proportionally much wider and more rounded than that of *M. campechiensis* and is more dorsoventrally compressed; and (4) the concentric lamellae of *M. hartae* are proportionally larger, more prominent, and more widely-separated than the closely-packed lamellae seen on *M. campechiensis*.

This new Georgian Subprovince venerid species is the ecological and morphological analogue of *Mercenaria texanum* (Dall, 1902) from the Texan Subprovince. Both endemic species are restricted to intertidal environments in coastal bays and inlets and both are smaller and more rounded than the sympatric *M. campechiensis*. Similarly, both *M. texanum* and *M. hartae* have smooth areas of reduced concentric sculpture. Of the known *Mercenaria* species of the Carolinian Province (discussed in Chapter 2), only *M. hartae* has the distinctive, highly polished semicircular patch bordering the shell margin.

Bibliography

Abbott, R.T. 1974. *American Seashells*. Second Edition. Van Nostrand Reinhold Company, New York. 663 pp.

Abbott, R.T. 1986. *Cantharus multangulus* new subspecies *grandanus* from northwest Florida (Buccinidae). *Nautilus* 100: 120–121.

Altena, C.O. Van R. 1971. On six species of Marine Mollusca from Suriname, four of which are new. *Zoologische Mededelingen* 45 (5): 75–86.

Briggs, J.C. 1974. *Marine Zoogeography*. McGraw-Hill Publishers, New York. 452 pp.

Briggs, J.C. 1995. *Global Biogeography*. Elsevier Press, Amsterdam. 452 pp.

Bright, T.J. and L.H. Pequegnat. 1974. *Biota of the West Flower Garden Bank*. Editors. Gulf Publishing Company, Houston, Texas. 285 pp.

Castelao, R.M. and J.A. Barth. 2006. Upwelling around Cabo Frio, Brazil: The importance of wind stress. *Geophysical Research Letters* 33: LO3602.

Coltro, J. 2004. New species of Conidae from Northeastern Brazil (Mollusca: Gastropoda). *Strombus* 11: 1–16.

Coltro, J. 2011. *Brazilian Conidae*. Femorale Publishers, Sao Paulo, Brazil. 93 pp.

Costa, P. M. and L. R. Simone. 1997. A new species of Conus Linne from the Brazilian Coast. *Siratus* 3(13): 3–6.

Crosse, H. 1876. Un nouvelle espece de *Voluta*. *Journal de Conchiologie* 24: 163–166, plate 5, figure 6.

Daccarett, E.Y. and V.S. Bossio. 2011. *Colombian Seashells from the Caribbean Sea*. L'Informatore Piceno. 384 pp.

Diaz, J.M. 1995. Zoogeography of Marine gastropods in the Southern Caribbean: A new look at provinciality. *Caribbean Journal of Science* 31 (1–2): 104–121.

Ekman, S. 1953. *Zoogeography of the Sea*. Sidgwick and Jackson Publishers, London. 417 pp.

Emiliani, C. 1955. Pleistocene temperatures. *Journal of Geology* 63 (6): 538–578.

Gonzalez-Rodriguez, E., J.L. Valentin, D.L. Andre, and S.A. Jacob. 1991. Upwelling and downwelling at Cabo Frio (Brazil): Comparison of biomass and primary productivity responses. *Journal of Plankton Research* 14 (2): 289–306.

Harasewych, M.G. and E.J. Petuch. 1980. *Sassia lewisi*, a new Cymatiid gastropod from the Caribbean Sea. *Nautilus* 94 (3): 121–122.

Johnston, A.K. 1856. *The Physical Atlas of Natural Phenomena*. W. Blackwood, Publisher, Edinburgh. 420 pp.

Leal, J. H. 1991. *Marine Prosobranch Gastropods from Oceanic Islands off Brazil*. W. Backhuys/UBS Publishers, Oegstgeest, The Netherlands.

Lopes, R.P. and L.R.L. Simone. 2012. New fossil records of Pleistocene Marine mollusks in Southern Brazil. *Revista Brasileira de Paleontologia* 15 (1): 49–56.

Marelli, D.C., W.G. Lyons, W.S. Arnold, and M.K. Krause. 1997. Subspecific status of *Argopecten irradians concentricus* (say, 1822) and of the Bay Scallops of Florida. *Nautilus* 110 (2): 42–44.

Matano, R.P., E.D. Palma, and A.R. Piola. 2010. The influence of the Brazil and Malvinas currents on the southwestern Atlantic shelf circulation. *Ocean Science Discussions* 7: 837–871.

Merle, D. and B. Garrigues. 2008. New Muricid species (Mollusca: Gastropoda) from French Guiana. *Zoosystema* 30 (2): 517–526.

Mikkelsen, P.M. and R. Bieler. 2007. *Seashells of Southern Florida: Living Marine Mollusks of the Florida Keys and Adjacent Regions: Bivalves*. Princeton University Press, Princeton, NJ. 496 pp.

Odum, E.P. 1971. *Fundamentals of Ecology*. Third Edition. W.B. Saunders, Philadelphia. 574 pp.

Okutani, T. 1982. A new genus and five new species of gastropods trawled off Suriname. *Venus* 41: 109–120.

Oleinik, A.E., E.J. Petuch, and W.C. Aley. 2012. Bathyal gastropods of Bimini Chain, Bahamas. *Proceedings of the Biological Society of Washington* 125 (1): 19–53.

Petuch, E.J. 1972. *Strombus costatus* and *Morum dennisoni* collected off the North Carolina Coast. *Veliger* 15: 32–35.

Petuch, E.J. 1974a. A new *Terebra* from the coral reef areas off North Carolina. *Veliger* 17 (2): 205–208.

Petuch, E.J. 1974b. Two new Pacific cone shells (Gastropoda: Conidae) and a new *Pleurotomella* from the Hatteras Abyssal Plain. *Veliger* 17 (3): 40–43.

Petuch, E.J. 1976. An unusual Molluscan assemblage from Venezuela. *Veliger* 18 (3): 322–325.

Petuch, E.J. 1979a. New gastropods from the Abrolhos Archipelago and Reef Complex. *Proceedings of the Biological Society of Washington* 92 (3): 510–526.

Petuch, E.J. 1979b. A new species of *Siphocypraea* (Gastropoda: Cypraeidae) from northern South America with notes on the genus in the Caribbean. *Bulletin of Marine Science* 29 (2): 216–225.

Petuch, E.J. 1980a. A new *Falsilyria* and a new *Conus* from Roatan Island, Honduras (Atlantic). *Nautilus* 94 (3): 115–118.

Petuch, E.J. 1980b. A new species of *Conus* from Southeastern Florida (Mollusca: Gastropoda). *Proceedings of the Biological Society of Washington* 93 (2): 299–302.

Petuch, E.J. 1981a. A Relict Caenogastropod Fauna from northern South America. *Malacologia* 20 (2): 307–334.

Petuch, E.J. 1981b. A Volutid species radiation from northern Honduras, with notes on the Honduran Caloosahatchian Secondary Relict Pocket. *Proceedings of the Biological Society of Washington* 94 (4): 1110–1130.

Petuch, E.J. 1982a. Geographical heterochrony: Contemporaneus coexistence of Neogene and recent Molluscan Faunas in the Americas. *Palaeogeography, Palaeoclimatology, and Palaeoecology* 37: 277–312.

Petuch, E.J. 1982b. Paraprovincialism: Remnants of paleoprovincial boundaries in recent Marine Molluscan provinces. *Proceedings of the Biological Society of Washington* 95 (4): 774–780.

Petuch, E.J. 1986a. The Austral-African Conid Subgenus *Floraconus* Iredale, 1930, taken off Bermuda (Gastropoda: Conidae). *Proceedings of the Biological Society of Washington* 99 (1): 15–16.

Petuch, E.J. 1986b. New South American gastropods in the genera *Conus* (Conidae) and *Latirus* (Fasciolariidae). *Proceedings of the Biological Society of Washington* 99 (1): 8–14.

Petuch, E.J. 1987. *New Caribbean molluscan faunas*. Coastal Education and Research Foundation, Charlottesville, VA. 158 pp.

Petuch, E.J. 1988. *Neogene History of Tropical American Mollusks*. Coastal Education and Research Foundation, Charlottesville, VA. 217 pp.

Petuch, E.J. 1991. A new molluscan faunule from the Caribbean Coast of Panama. *Nautilus* 104 (2): 57–71.

Petuch, E.J. 1992a. Molluscan discoveries from the tropical western Atlantic region. Part 1. New species of *Conus* from the Bahamas Platform, Central American and northern South American Coasts, and the Lesser Antilles. *La Conchiglia* 23 (264): 36–40.

Petuch, E.J. 1992b. Molluscan discoveries from the tropical western Atlantic region. Part 2. New species of *Conus* from the Bahamas Platform, Central American and northern South American Coasts, and the Lesser Antilles. *La Conchiglia* 24 (265): 10–15.

Petuch, E.J. 1992c. New mollusks from Los Roques Archipelago, Venezuela, an isolated Caribbean atoll. *La Conchiglia* 23 (262): 5–11.

Petuch, E.J. 1993a. Molluscan discoveries from the tropical western Atlantic. Part 2. A new species of *Leporiconus* Iredale, 1930 from the San Blas Islands, Panama. *La Conchiglia* 24 (266): 57–59.

Petuch, E.J. 1993b. A new *Polystira*. *La Conchiglia* 24 (267): 62–63.

Petuch, E.J. 1994. *Atlas of Florida Fossil Shells (Pliocene and Pleistocene Marine Gastropods)*. Graves Museum of Archaeology and Natural History, Dania, FL. 394 pp.

Petuch, E.J. 1995a. Molluscan discoveries from the tropical western Atlantic region. *La Conchiglia* 27 (275): 36–41.

Petuch, E.J. 1995b. Molluscan diversity in the late Neogene of Florida: Evidence for a two-staged mass extinction. *Science* (13 October) 270: 275–277.

Petuch, E.J. 1997. *Coastal Paleoceanography of Eastern North America*. Kendall/Hunt Publishing Company, Dubuque, IA. 373 pp.

Petuch, E.J. 1998a. Molluscan discoveries from the tropical western Atlantic region. Part 5. New species of *Conus* from the Bahamas, Honduran Banks, San Blas Archipelago, and northeastern South America. *La Conchiglia* 30(287): 25–36, 62.

Petuch, E.J. 1998b. The Molluscan fauna of the Wawa River Region, Miskito Coast, Nicaragua: Ecology, biogeographical implications, and descriptions of new taxa. *Nautilus* 111 (1): 22–44.

Petuch, E.J. 2000. A review of the Conid Subgenus *Purpuriconus* da Motta, 1991 with the descriptions of two new Bahamian species. *Ruthenica* 10: 81–87.

Petuch, E.J. 2001. New gastropods named for Frederick M. Bayer, in recognition of his contributions to tropical western Atlantic malacology. *Proceedings of the Biological Society of Washington* 10: 334–343.

Petuch, E.J. 2002. New deep water gastropods from the Bimini Shelf, Bimini Chain, Bahamas. *Ruthenica* 12 (1): 59–72.

Petuch, E.J. 2004. *Cenozoic Seas: The view from eastern North America*. CRC Press, Boca Raton, FL. 308 pp.

Petuch, E.J. 2008. *The Geology of the Florida Keys and Everglades*. Thomson Publishers, Mason, Ohio. 84 pp.

Petuch, E.J. 2012. Miocene asteroid impacts: Proposed effects on the extinction patterns of eastern North American gastropods. In *Earth and Life: Global Biodiversity, Extinction Intervals, and Biogeographic Perturbations through Time*. J.A. Talent, Editor. Springer Publishers, Dordrecht, Netherlands, pp. 967–981.

Petuch, E.J. and M. Drolshagen. 2009. *Molluscan Paleontology of the Chesapeake Miocene*. CRC Press, Boca Raton, FL. 160 pp.

Petuch, E.J. and M. Drolshagen. 2011. *Compendium of Florida Fossil Shells, Volume 1. (Middle Miocene to Late Pleistocene Marine Gastropods; Families Strombidae, Cypraeidae, Ovulidae, Eocypraeidae, Triviidae, Conidae, and Conilithidae)*. MdM Publishers, Wellington, FL. 412 pp.

Petuch, E.J. and C.E. Roberts. 2007. *The Geology of the Everglades and Adjacent Areas*. CRC Press, Boca Raton, FL. 212 pp.

Petuch, E.J. and D.M. Sargent. 1986. *Atlas of the Living Olive Shells of the World*. Coastal Education and Research Foundation, Charlottesville, VA. 253 pp.

Petuch, E.J. and D.M. Sargent. 2011a. A new member of the *Gradiconus* species complex (Gastropoda: Conidae) of the Florida Keys. *Visaya* 3 (4): 98–104.

Petuch, E.J. and D.M. Sargent. 2011b. New species of Conidae and Conilithidae (Gastropoda) from the tropical Americas and Philippines, with notes on some poorly-known Floridian species. *Visaya* 3 (3): 37–58.

Petuch, E.J. and D.M. Sargent. 2011c. *Rare and Unusual Shells of the Florida Keys and Adjacent Areas*. MdM Publishers, Wellington, FL. 158 pp.

Petuch, E.J. and D.M. Sargent. 2012. *Rare and Unusual Shells of Southern Florida*. Conch Republic Books, Mount Dora, FL. 189 pp.

Poppe, G. and Y. Goto. 1992. *Volutes*. Mostra Mondiale Malacologia, L'Informatore Piceno. 348 pp.

Rezak, R., T.J. Bright, and D.W. McGrail. 1985. *Reefs and Banks of the Northwestern Gulf of Mexico: Their Geological, Biological, and Physical Dynamics.* Wiley Publishers, New York. 449 pp.

Rios, E. de C. 1994. *Seashells of Brazil.* Second Edition. Fundacao Universidade do Rio Grande, Publishers, Rio Grande do Sul, Brazil. 368 pp., 113 plates.

Tucker, J.K. 1994. The Crown Conch (*Melongena*: Melongenidae) in Florida and Alabama with the description of *Melongena sprucecreekensis* n. sp. *Bulletin of the Florida Museum of Natural History, Biological Sciences* 36 (7): 181–203.

Tucker, J.K. 2004. Catalog of recent and fossil turrids (Mollusca: Gastropoda). *Zootaxa* 682: 1–1295.

Tucker, J.K. and M.J. Tenorio. 2009. *Systematic Classification of Recent and Fossil Conoidean Gastropods.* Conchbooks, Hackenheim, Germany. 296 pp.

Tunnel, J.W., J. Andrews, N.C. Barrera, and F. Moretzsohn. 2010. *Encyclopedia of Texas Seashells: Identification, Ecology, Distribution, and History.* Texas A&M University Press, College Station. 512 pp.

Valentine, J.W. 1973. *Evolutionary Paleoecology of the Marine Biosphere.* Prentice-Hall, Englewood, NJ. 511 pp.

Vermeij, G.J. 1978. *Biogeography and Adaptation.* Harvard University Press, Cambridge, MA. 332 pp.

Vermeij, G.J. and E.J. Petuch. 1986. Differential extinction in tropical American mollusks: Endemism, architecture, and the Panama Land Bridge. *Malacologia* 27 (1): 29–41.

Vokes, E. 1964. The genus *Turbinella* (Mollusca: Gastropoda) in the New World. *Tulane Studies in Geology* 2(2): 39–68.

Waller, T.R. 2011. Neogene paleontology of the Northern Dominican Republic. 24. Propeamussidae and Pectinidae (Mollusca: Bivalvia: Pectinoidea) of the Cibao Valley. *Bulletins of American Paleontology* 381: 1–198.

Warmke, G. and R.T. Abbott. 1962. *Caribbean Seashells.* Livingston Publishing Company, Narberth, PA. 348 pp.

Weisbord, N.E. 1962. Late Cenozoic gastropods from northern Venezuela. *Bulletins of American Paleontology* 42 (193): 7–672.

Woodward, S.P. 1856. *Manual of Mollusca: or a Rudimentary Treatise of Recent and Fossil Shells.* John Weale, London. 484 pp.

Index

T - #0396 - 101024 - C256 - 254/178/14 - PB - 9781138033757 - Gloss Lamination